CAT BEHAVIOR AND TRAINING
Veterinary Advice for Owners

COMPILED BY: LOWELL ACKERMAN, DVM
EDITED BY: LOWELL ACKERMAN, DVM
GARY LANDSBERG, DVM
WAYNE HUNTHAUSEN, DVM

Publisher's Note: *The portrayal of pet products in this book is strictly for instructive value only; the appearance of such products does not necessarily constitute an endorsement by the editor, authors, the publisher, or the owners of the cats portrayed in this book.*

Distributed in the UNITED STATES to the Pet Trade by T.F.H. Publications, Inc., One T.F.H. Plaza, Neptune City, NJ 07753; distributed in the UNITED STATES to the Bookstore and Library Trade by National Book Network, Inc. 4720 Boston Way, Lanham MD 20706; in CANADA to the Pet Trade by H & L Pet Supplies Inc., 27 Kingston Crescent, Kitchener, Ontario N2B 2T6; Rolf C. Hagen Inc., 3225 Sartelon St. Laurent-Montreal Quebec H4R 1E8; in CANADA to the Book Trade by Vanwell Publishing Ltd., 1 Northrup Crescent, St. Catharines, Ontario L2M 6P5 ; in ENGLAND by T.F.H. Publications, PO Box 15, Waterlooville PO7 6BQ; in AUSTRALIA AND THE SOUTH PACIFIC by T.F.H. (Australia), Pty. Ltd., Box 149, Brookvale 2100 N.S.W., Australia; in NEW ZEALAND by Brooklands Aquarium Ltd. 5 McGiven Drive, New Plymouth, RD1 New Zealand; in Japan by T.F.H. Publications, Japan—Jiro Tsuda, 10-12-3 Ohjidai, Sakura, Chiba 285, Japan; in SOUTH AFRICA by Lopis (Pty) Ltd., P.O. Box 39127, Booysens, 2016, Johannesburg, South Africa. Published by T.F.H. Publications, Inc.

MANUFACTURED IN THE UNITED STATES OF AMERICA
BY T.F.H. PUBLICATIONS, INC.

Contents

About the Editors

Dr. Lowell Ackerman is a Diplomate of the American College of Veterinary Dermatology and a consultant in the fields of dermatology, nutrition and medicine. He is the author of 13 books and over 150 articles and book chapters dealing with pet health care and is the co-author of a veterinary textbook dealing with pet behavior problems. In addition, he has lectured extensively on these subjects on an international basis. Dr. Ackerman is a member of the American Veterinary Society of Animal Behavior.

Dr. Gary Landsberg is a companion animal veterinarian at the Doncaster Animal Clinic in Thornhill, Ontario, Canada. He is also extremely active in the field of pet behavior and offers a referral consulting service for pets with behavior problems. He is the co-author of two veterinary textbooks dealing with pet behavior problems. Dr. Landsberg has lectured throughout North America and Europe and is the past president of the American Veterinary Society of Animal Behavior.

Dr. Wayne Hunthausen is a pet behavior therapist who works with people, their pets and veterinarians throughout North America to help solve pet behavior problems. He writes for a variety of veterinary and pet publications and is co-author of two textbooks dealing with pet behavior. Dr. Hunthausen frequently lectures on pet behavior and currently serves on the behavior advisory board for several scientific journals. He is currently president of the American Veterinary Society of Animal Behavior.

Facing Page: Throughout the ages, cats have fascinated their human companions with their beauty and unique personalities. Photo by Robert Pearcy.

International Directory of Behavior Organizations

Listed below are some of the professional organizations that are involved in the training and advancement of the field of cat behavior.

American Veterinary Society of Animal Behavior
c/o Secretary-Treasurer
Dr. Debra Horwitz
Veterinary Behavior Consultations
253 S. Graeser Road
St. Louis, MO 63141

American College of Veterinary Behaviorists
c/o Dr. Katherine Houpt, Secretary
Dept. Of Physiology
College of Veterinary Medicine
Cornell University
Ithaca, NY, 14853-6401

Companion Animal Behavior Therapy Study Group
Mr. D. Mills, Secretary
De Montford University, Lincoln
Caythorpe Court
Catythorpe
Nr Grantham, Lincs, NG32 3EP, UK

International Society for Animal Ethology
Dr. S.M. Rutter, ISAE Membership
Institute of Grassland and Environmental Research
North Wyke, Okehampton, Devon, EX20 2SB, UK

Animal Behavior Society
Dr. Janis Driscoll
Dept. Of Physiology
Campus P.O. Box, 173364
Denver, CO, 80217-3364, USA

European Society for Veterinary Clinical Ethology
Dr. J. Dehasse
129, Avenue de la Fauconnerie
92, B-1170
Brussels, Belgium

Facing Page: An attractive agouti-patterned kitten photographed by Robert Pearcy.

Preface

Behavioral problems are the number one killer of pets in this country. This might seem surprising. Surely behavioral problems can't be the number-one killer of pets. It must be car accidents, or perhaps cancer. The sad truth is that as many as eight million pets are euthanized (killed) each year because of behavior problems. Veterinarians spend about 20% of their time discussing behavior concerns with clients. Between 50% and 70% of animals in shelters are there because their owners either couldn't or wouldn't deal with their behavior problems. If these behavior problems were caused by a virus, we would consider it an epidemic. Make no mistake about it—unacceptable behavior, not disease, is the number one killer of pets in this country.

What can we do about it? Knowledge is the key! Sound, timely and practical information is the best defense. Proper socialization and training are the best way to prevent problems from occurring. Early intervention with behavior counseling results in fewer problems that progress to a point where the animal is abandoned or destroyed. This book bridges the gap between understanding the causes of behavior problems, preventing them, and managing them once they're already evident. This information is relayed by some of the pre-eminent experts in this relatively new field.

Only a veterinarian is trained and legally permitted to diagnose and treat medical problems. Thus, the veterinarian is the first person to be consulted with a behavior

Facing Page: Establishing "house rules" during kittenhood will help to ensure a successful training program. Persian kitten photographed by Robert Pearcy.

A non-purebred cat can make as equally fine a pet as can a purebred cat.

problem. Before any condition is dismissed as a "training failure," it is critical to determine that there is not a medical reason for the problem. In addition, some problems are more amenable to pharmacological or surgical treatment than to behavioral modification and training. Should your veterinarian feel that a referral to a veterinary specialist or other behavior counselor is warranted, he or she can refer you to the most appropriate person in your area for your pet's problem.

As our knowledge of animal behavior grows, so does the way we train and shape our pet's behavior. Physical discipline and harsh punishment are seldom indicated and not only are inhumane but also can lead to excessive fear and anxiety. Reward, motivational training, behavior modification techniques and a firm but fair approach to discipline are the order of the day.

This book provides a wealth of information for the pet-loving public. Written entirely by veterinarians, it provides insight into the effectiveness of various treatments for the most common behavioral problems and how those problems are best prevented.

Lowell Ackerman, DVM, PhD
Gary Landsberg, B.Sc., DVM
Wayne Hunthausen, DVM

Facing page: American Shorthair, silver tabby. Photo by Robert Pearcy.

Dr. Debra Horwitz graduated from Michigan State University, College of Veterinary Medicine, in 1975. Since 1982, Dr. Horwitz has had a referral practice for behavioral problems in companion animals. From 1990 onward, she has devoted her practice energies exclusively to the practice of behavioral medicine in companion animals. Her practice is located in St. Louis, Missouri where she is now affiliated with Associated Veterinary Specialists. She is a frequent lecturer, both nationally and internationally. Her articles on companion animal behavior appear in both veterinary and popular publications. She appears locally in St. Louis on both television and radio as well as in local print publications and lectures.

Basic Feline Behavior and Social Development

by Dr. Debra Horwitz
Veterinary Behavior Consultations
12462G Natural Bridge Road
Bridgeton, MO 63044

INTRODUCTION

Cats have been around since antiquity, as objects of worship, derision and as pets. Throughout history, cats have had roles in society, ranging from their association with the gods and goddesses of Egypt, to being identified with witchcraft. The past 10-15 years have shown a steady increase in the numbers of cats as household pets. With the increase of cats as pets, more information has been gathered on their social development and behavior.

In the past, cats have been thought of as asocial. This does not mean that cats are anti-social, rather it means that cats tend to be solitary. The early interaction between people and cats was of the utilitarian type, i.e., cats were kept around for their rodent control ability. In fact, some people say that the cat actually domesticated itself! Over time, the relationship between people and cats has changed. As more cats have become house pets, they have been selected for greater sociability. Yet each cat retains a little of its "wild" ways, and it is this mystery that has enthralled the cat-owning public for centuries.

THE SOCIAL NATURE OF THE CAT

As mentioned above, cats are more adapted to a solitary lifestyle than are dogs. Yet, social interactions among female cats and neutered males are quite common. Depending on a variety

"Now what do I do?" Like many other young household pets, kittens can get themselves into any number of predicaments.

Above: Play behavior. Social interactions among female cats and neutered males are quite common.
Below: Cat on the prowl. Cats that are allowed to roam at will can be exposed to a number of dangers.

of factors including genetics, early socialization, sex, availability of food and space limitations, most cats seem to adapt extremely well to living with other cats and people. Most enjoy human companionship and will seek out their owners for affection. Yet, unlike the dog, the cat is not motivated to "please" the owner. It is this trait that can both endear them to owners, or cause them to lose their homes.

FELINE COMMUNICATION

Cats communicate by using body postures, vocalizations, and marking behaviors. Due to the solitary nature of the cat, body postures do not play as much of a role as they do in the dog. A few of the body postures used by cats will be mentioned here. Most of the cat's body postures tend to be "go away" or defensive body postures such as the Halloween cat posture. These body postures are often associated with piloerection (the standing on end of the hair). Friendly approaches are usually characterized by a cat with the tail held vertically, the hair flat. Play postures, like rolling over on the back, are also seen in cats. Submissive postures, which are used infrequently, consist of a crouch with the ears flat against the head.

Cats use marking behavior extensively to communicate. Cats can mark with urine, stool, their claws and scent glands on their feet, face and tail. Many owners have seen their cats rub their face, chin and feet on household objects. These areas of the body have sebaceous glands that most likely produce a scent. It is thought that when a cat rubs, or is petted in one of these areas, an odor is left behind. Perhaps when our cats rub on us, they are marking us as property. Marking is an important behavior for the cat, yet at the same time can be a problem behavior for the owner.

SOCIALIZATION

The importance of the "sensitive period" of socialization in domestic animals has been emphasized time and again in both scientific and popular literature. A sensitive period is the time when an experience that a young animal does or does not have will have the greatest effect on its later behavior. The time period that most pet owners are concerned with is the socialization period. It is this time period which helps shape the later behavior of the pet in its relationship to people. At the present time, the sensitive period for socialization in the cat is thought to be from 3-7 weeks, with a range of 2-9 weeks. The early onset of this period makes the first few weeks of the kitten's life very important.

To aid in proper socialization, it is essential that kittens be gently handled early and frequently in the first few weeks of life. This early handling helps the kittens in their transition from life in a litter to life as pets. Research on kittens has

Above: A friendly approach. Note that the tail is held vertically and that the hair on the back lies flat. *Below:* This cat is exhibiting an antagonistic posture. Its ears are pivoted toward the sides of its head.

Two types of cat behavior. One cat is using a tree limb to hone his claws; the other is about to spray a tree limb with urine. The latter behavior is known as scent-marking.

shown that early handling of kittens decreased their approach time to strangers and increased the amount of time that they stayed with them. Early handling also increases the likelihood that the kitten will relate well to people when placed in a home after weaning at 6-9 weeks of age. Handling the young kitten also has the benefit of accelerated motor and nerve development.

Kittens are also more easily socialized to other species during the socialization period. If raised with other animals normally considered to be prey or enemies, they will get along with them quite well as companions. Young kittens introduced to other animals can and will live with them quite peacefully. It is important to understand that during the sensitive period attachments are most easily formed. This does not mean that attachments do not form at other times, it just may take longer to establish them. Certainly, kittens should not be adopted earlier than 6-8 weeks, after they are weaned from their mother. However, it is important that sufficient early handling take place during the first weeks to facilitate the transition to the kitten's future home. Through early handling, kittens become more accepting of people, which helps to create calmer, friendlier pets.

CAT PERSONALITIES

Studies undertaken about cats have shown that it may be possible to classify cat "personalities" in the same way as has been done for dogs. One such study was interested in trying to identify cats that were shy, timid or fearful and those that were confident. This study found that timid cats took significantly longer to approach people and be held by them. Another study identified cats that were "shy" and those that were "trusting". That research noted that trusting cats were trusting regardless of where they encountered people, while shy cats were more fearful the further from home they were encountered. The results of these two studies and one other have come up with two common cat personality types: (a) sociable, confident and easygoing; (b) timid, shy and unfriendly. Additional research has also indicated an active aggressive personality type. An offshoot of studying personality typing is trying to determine what influences the development of personality type. Recent information seems to indicate that the friendliness of cats may be partly influenced by the father.

EARLY LEARNING

In the early weeks of a kitten's life, it learns a great deal from the

The best of friends. Cats and dogs can get along well together if they are introduced at an early age. Photo by Robert Pearcy.

mother, or queen. Cats are very good at observational learning, and the queen starts to teach her kittens at a young age. Through observation of the queen, kittens learn elimination behaviors and predatory behavior.

Kittens will begin to spend time in the litter box at about 30 days of age. They will learn appropriate litter usage through observation of the queen and certain odors associated with the litter box. The type of litter material that is used at this time may influence litter choices for the kitten at a later date. In other words, a kitten that was eliminating outdoors in the dirt may prefer that substance over clay litter.

If allowed access to prey, the queen will begin to bring them to her offspring at about 32-36 days of age. She will release them for the kittens and monitor their responses and hunting techniques. Because of this early experience, the prey that an adult cat hunts is usually the same as the prey hunted by the queen. The kittens will also begin to wean and eat solid food at about 4-5 weeks. The choice of food can be influenced by the queen. The kittens will taste her food and may learn to prefer it over another choice.

ACCLIMATION

Once a kitten has been chosen, it is time to take the kitten home. Cats and kittens should always be transported in some kind of carrier in the car; even a box will suffice at this young age. Upon arriving at the new home, place the kitten in a small quiet area with food and a

In the early weeks of a kitten's life, it learns a great deal from its mother, or queen. American Shorthairs photographed by Robert Pearcy.

Mutual grooming. This behavior can be observed frequently in households that have more than one cat.

Longhaired breeds require a regular grooming regimen to keep them looking their best.

litter box. If the kitten is very tiny, a small litter box with lowered sides may be necessary at first. It can be helpful to duplicate the type of litter material used in the previous home. Be sure to inspect this confinement area at floor level for hiding places behind things and in nooks and crannies. No matter how outgoing the kitten, it may be scared and may look for a safe secluded area to hide.

Cats and kittens do have a need to investigate their new surroundings. By limiting the space initially, you make this a more manageable task for a young kitten. It is essential to always monitor a young kitten's investigations. It appears that cats use a random method of search in strange areas and that there is a high degree of trial-and-error learning. After giving the kitten some quiet time in the first few days, a new owner can slowly allow the kitten increased access to other areas of the home.

If there are already additional pets in the home, care must be taken to protect the new kitten from possible aggression from resident pets. Slowly allow the current pets to meet the new kitten. This is best done first through a closed door for several days to acclimate the residents to the smell of the kitten. Then the existing pets can be introduced to the new kitten at a distance. It is often helpful to use food treats to make this first encounter a pleasurable experience. Have the animals meet under supervision

"We were here first." This mother cat and her youngster stake their claim to a comfy chair.

several times before assuming that they get along. Once a relationship is established, there still may be times when separation is needed. Young animals are very energetic and can be annoying to older animals who are not interested in playing.

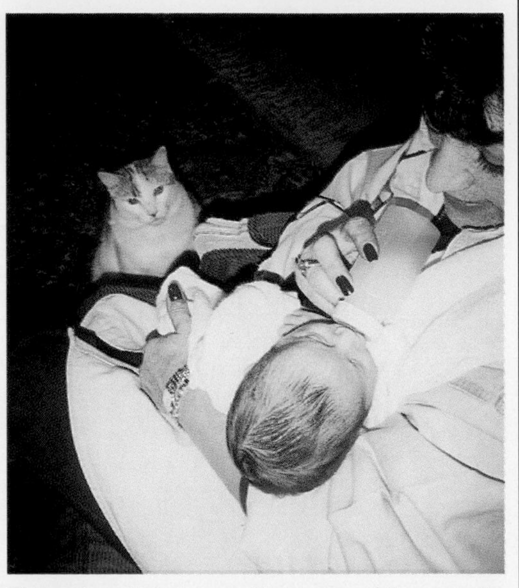

The series of photos on this page and the opposite page documents the reactions of a cat to the new human member of its family. The highly curious nature of the cat is readily apparent in all of the photos. Photos courtesy of Dr. Burt Frank.

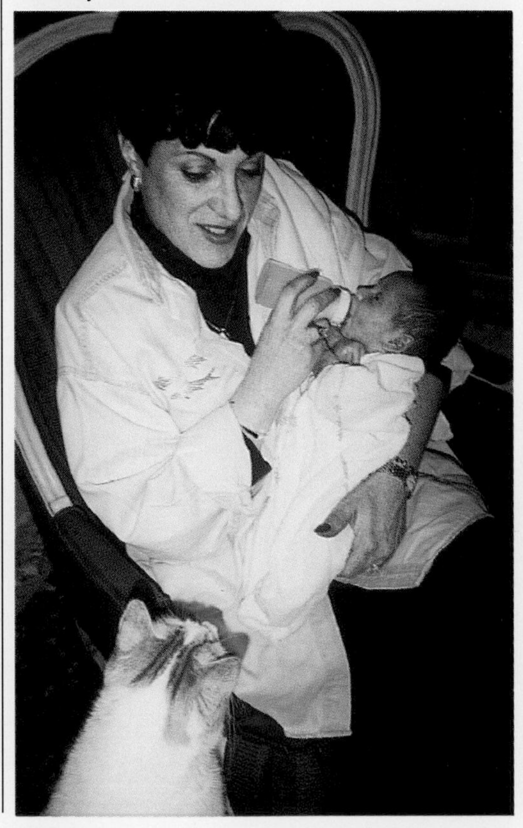

HOME SOCIALIZATION

Kittens are natural explorers and will use their claws to climb up onto almost anything. In the first few weeks, slow access to the home will allow exploration, as well as the ability to monitor the kitten's behavior. During this time, it is important to allow adequate stimulating play. Stalking and pouncing behaviors are important play behaviors in kittens. They have an essential role in neural and muscular development. If given sufficient outlet for these behaviors with toys and playtime, kittens will be less likely to use people for these activities. Toys for kittens need to be of the interactive type, i.e. toys that move and are light. They can include wads of paper, kitten fishing rods, walnuts, and small balls or stuffed toys. Caution should be exercised to eliminate toys that are too heavy for the kitten to move with their paws, too large to pick up in their mouths or so small as to be swallowed. It is also best not to use toys that encourage kittens to jump and grab at the owner's hands. This only encourages wild play that can end up injuring owners. Unless a kitten is allowed acceptable outlets for play, this

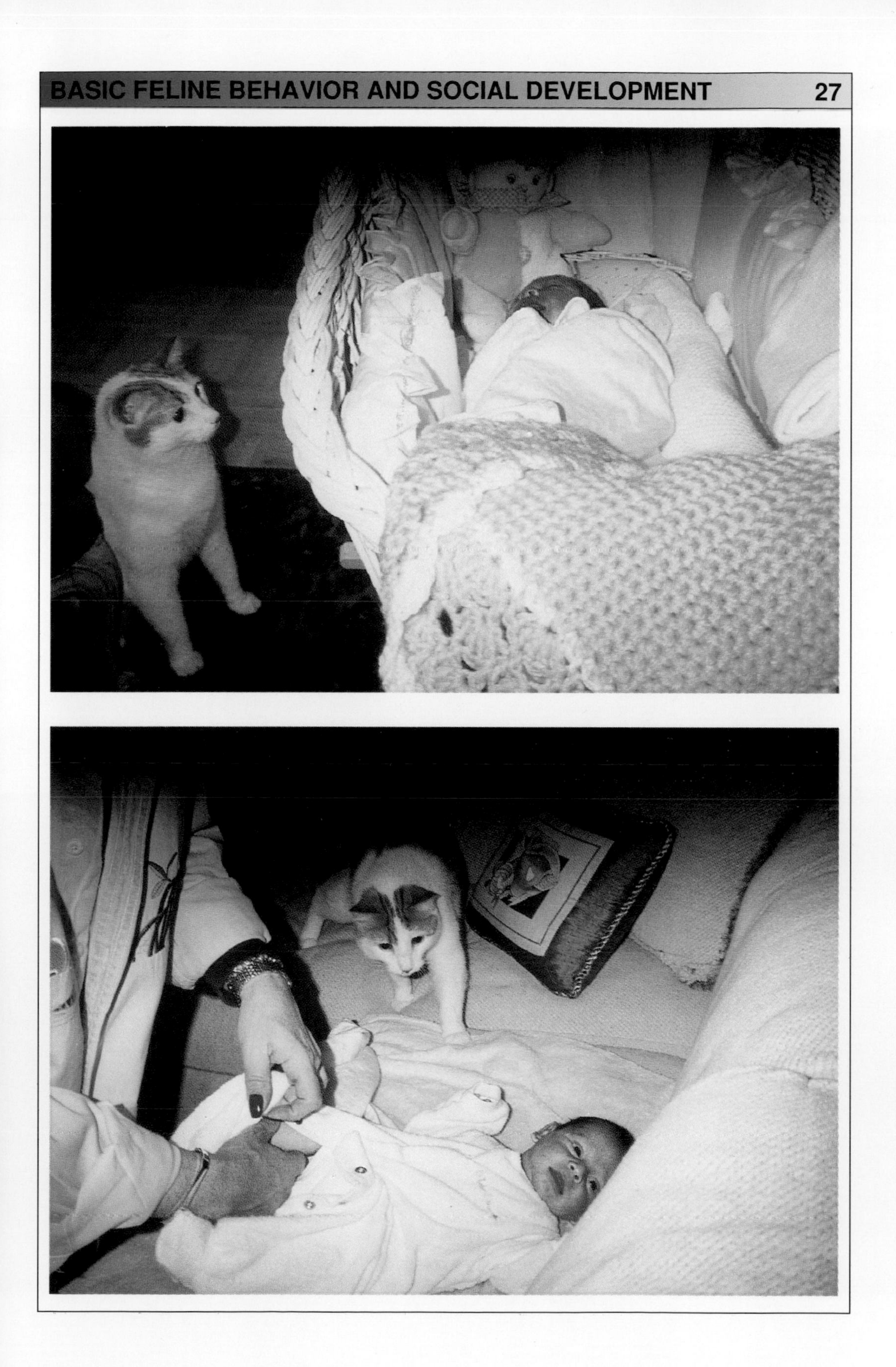

This page and opposite page: Continuation of the series of photos documenting a cat's reactions to a new baby. Photos courtesy of Dr. Burt Frank.

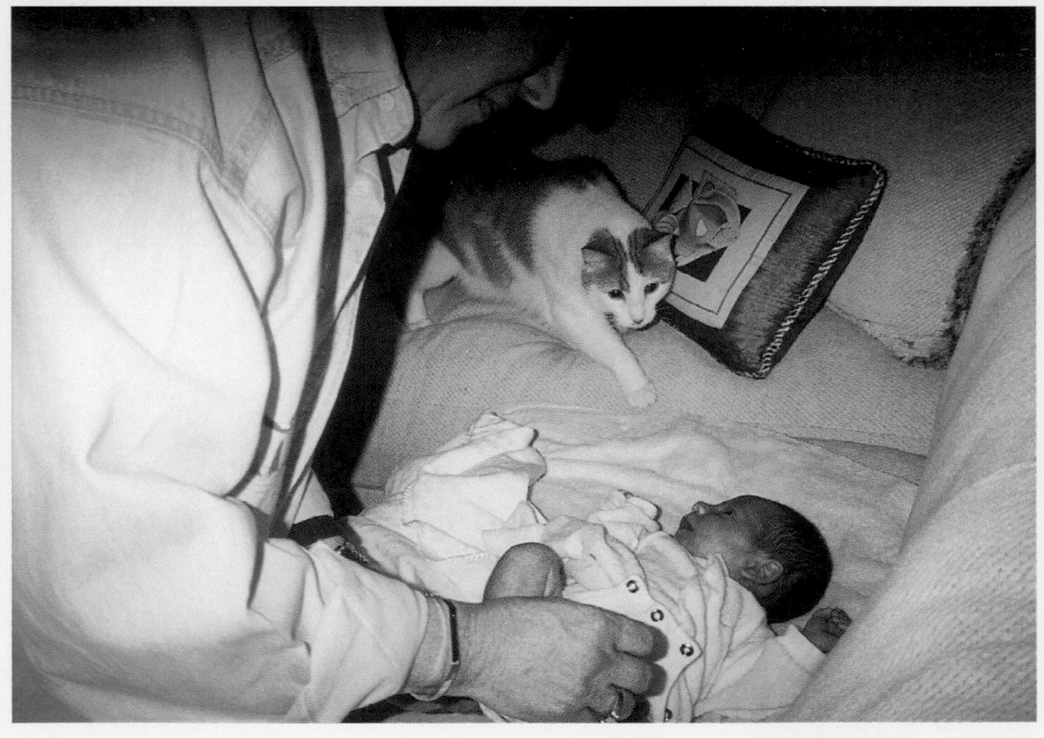

excessive energy can become a problem behavior.

Discipline of a young inquisitive kitten may be necessary to protect property and people. Harsh physical discipline is not necessary in young animals and can have the result of making the kitten afraid of people. Noise is an effective deterrent to kittens and can intimidate them enough to inhibit behavior. This can be accomplished by hand clapping, shaker cans or air horns. Even hissing at a young kitten can be intimidating. Another deterrent is the use of "remote punishment". Remote punishment consists of using something to stop the behavior that appears unconnected to the punisher. Examples are sprayer bottles, noise makers, and other loud noises. Remote punishment serves to remove the association of the punishment from the person doing it and place it on the action being performed at the time. Hopefully this will carry over to other times when the owner is not present and the cat will inhibit the undesirable behavior on its own.

Sometimes young kittens can be so rambunctious when left alone that it may be necessary to place them in a "kitten-proof" room. This room should have lots of toys, boxes to climb in, lightweight objects to hit and climbing towers. This will help channel a kitten's natural energy in the right direction. Because cats are nocturnal to some degree, this may be a good spot for a kitten to spend the night.

MULTIPLE-CAT HOUSEHOLDS

Another area of cat ownership that has received sparse practical attention over the years is the spatial requirements of cats. Also of interest is cat sociability in regard to the numbers of cats kept in the home. Information gathered from studying the ranges of farm cats shows that the ranges of cats are quite variable in size. The range size depends on factors including, but not limited to, food, breeding cycle, and whether the cats are related. What does appear to be important is that male cats have larger ranges than female cats. One could possibly infer that, in general, female cats adapt better to indoor living than do male cats. Although cats seem to adapt to variously sized home ranges, the diversity of that area is important to provide stimulation and interest. Therefore, it is important, especially if the space is small, that it be diverse, i.e. quality vs. quantity. One means to accomplish diversity is to provide a variety of heights, especially for resting places. Cats particularly like to arrange themselves vertically within the space that they have. This is why cats like to climb up on bookshelves, refrigerators, and mantels! A cat will also need an assortment of play objects to remain stimulated. These can be boxes, bags, kitty condos—anything to relieve the monotony that may contribute to stress and behavioral problems.

Play time. Many people keep a pair of cats so that the animals can provide companionship for each other when their owner is not at home. Photo by Robert Pearcy.

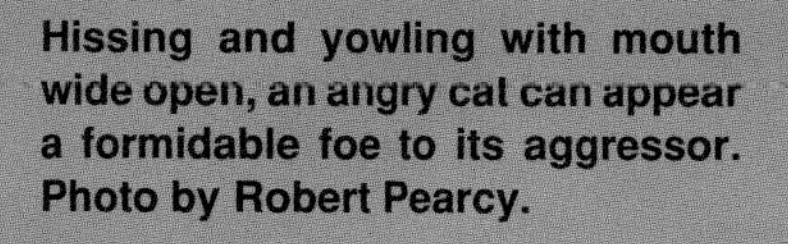

Hissing and yowling with mouth wide open, an angry cat can appear a formidable foe to its aggressor. Photo by Robert Pearcy.

This cat and its canine companion are both of undetermined lineage. Photo by Robert Pearcy.

Besides the diversity of the living area, the number of other cats in the environment influences social relationships. It has been shown that cats can vary from those that are sociable, i.e., enjoy contact and those that are solitary. Much of this sociability is influenced by the cat's early environment and interaction with other cats and kittens during the sensitive period of life. Genetics are also involved, as it is believed that certain aspects of social behavior are inherited. Research suggests that cats do enjoy companionship and house-mates if they are of the social type, while others that are solitary prefer to be an "only" cat. The best time to obtain two cats would be to obtain two kittens from the same litter and allow them to grow up together. If over time one of the housemates passes away, it is not always desirable or even necessary to replace that cat. Just because a cat was raised with another cat, it does not automatically mean that the remaining cat will get along with other cats. Each cat is an individual, and its personality must be taken into consideration when adding additional cats to the home.

Russian Blue photographed by Robert Pearcy.

It is essential to remember that because of spatial considerations, there is definitely a threshold for the number of cats in a household. When a home reaches that point, behavioral problems will be manifested. Among those problems

are housesoiling, with urine and/or stool, spraying, vocalization, and aggression.

In households that have multiple cats, it may be necessary to provide some cats with "alone time." This is a period of time in which the cat has access to food, water, litter box and resting areas without being disturbed by other cats. This can be especially important for cats that are outcasts in the group or for cats that seem stressed by having many other cats in the environment. Stressed cats will often exhibit inappropriate behavior such as marking with urine or stool, or hiding or vocalizing.

EFFECTS OF IMPROPER SOCIALIZATION

Lastly, improper socialization can contribute to the development of behavioral problems that can spoil the bond between human and animal. While socialization is important, certain genetic influences also play an important role in behavior. An additional factor is the individuality of each cat and how it responds to socialization. Common problems with kittens are misdirected play aggression and destruction due to scratching behaviors. With adult cats, problems can be aggression toward people, redirected aggression towards housemates or people due to external or internal stimuli, and housesoiling. Often the answers to these problems lie in adequate play alternatives for play aggression and proper placement and training to a scratching post for clawing behaviors. Proper socialization with people and other animals will often help to prevent aggression, and adequate litter maintenance plays a role in preventing housesoiling.

Another area of socialization that can create problems is the hand-raised kitten who lacks proper input from other cats and also can be aggressive toward non-owners.

Facing Page: Aggressive behavior in cats can be manifested in a number of different ways. Photo by Robert Pearcy.

SUMMARY

Throughout time, felines have fascinated and captivated people with their beauty and their special personalities. By understanding their growth, development and their needs we can make cat ownership a pleasurable experience.

ADDITIONAL READING

Beaver, Bonnie. *Feline Behavior: A Guide for Veterinarians,* W. B. Saunders, Philadelphia, PA 1992.

Clark, Ross., *Medical, Genetic & Behavioral Aspects of Purebred Cats,* Forum Publications, Fairway, Kansas. 1992.

Turner, Dennis C. Bateson, Patrick. *The Domestic Cat,* the Biology of Its Behaviour. Cambridge University Press, New York, 1988.

Dr. Ramsay Macdonald, a graduate of the Ontario Veterinary College (1959), has been a small animal practitioner for 35 years in the Toronto area. The owner and director of an AAHA-accredited veterinary hospital and a vaccination service, he has had a special interest in animal behavior since the late 1960s and is a member of the American Veterinary Society of Animal Behavior. Dr. Macdonald was the President of the Toronto Academy of Veterinary Medicine in 1973-1974 and a founding member of the Ontario Veterinary Medical Association.

Selecting the Perfect Cat for the Family

By Ramsay Macdonald, DVM
Sheppard Avenue Veterinary Hospital
4023 Sheppard Avenue
Agincourt, Ontario, Canada

INTRODUCTION

Selecting an appropriate cat for the family is critical since cats are long-lived and may spend up to twenty years (sometimes more!) in your home. That's why it is often surprising that people buy a cat on impulse and hope it fits in with their lifestyle. Unfortunately, about eight million cats are killed each year in the United States and Canada for non-medical reasons. Therefore, there must be something wrong with the current method of choosing pet cats. Let's examine some of the criteria that should figure into our selection process.

SELECTION CRITERIA

Age

Everyone loves a kitten. Although some breeders prefer to keep their kittens until 12 weeks old for various reasons, kittens must receive contact and exposure to people and other species by 7-9 weeks of age to ensure proper socialization. If cats don't socialize with people and other animals by that time, there is a good possibility that they will never become "social" pets. This means that kittens would have to be adopted by seven weeks of age to ensure proper relationships with all family members and other pets in the home. On the other hand, to develop healthy social skills with other cats, it would be ideal to keep a kitten with its mother and littermates. The best solution to this dilemma is to:

- obtain a kitten at around seven weeks of age or older that has had sufficient contact and handling by

"Up, Up, and Away?" Cats love to play with objects that they can easily manipulate.

people with other pets (e.g., dogs, birds, etc.), or;

- obtain a kitten before seven weeks of age and ensure ongoing cat-to-cat interaction with an existing cat in the household, or;
- adopt two kittens so there is ongoing socialization with another cat.

Of course, there's nothing wrong with adopting an adult cat, and there are some major advantages. Cats approach adulthood by six months of age but expect physical growth to continue until ten months old. When you adopt an adult cat, you have an excellent chance to see how well it was socialized, and how well it relates to you and other household pets. The adult feline that was properly cared for is probably already fully vaccinated and, perhaps, neutered. The prime benefit of the adult cat is that there are fewer surprises—what you see is what you get! That can mean a lot.

Cats approach adulthood by six months of age but physical development continues until the age of ten months.

Male or Female?

The most obvious advantage of selecting a female is her capability of producing a litter of kittens. However, along with that is her tendency to do this over and over again. Cats generally have their kittens in the spring and fall, so that just before these seasons, any sexually mature cat (over five months of age) may begin to exhibit overt signs of "heat." Cycles may differ greatly between cats: signs of heat can appear anywhere from two weeks to a few months apart. During heat, female cats attempt to attract or seek out males by using loud vocalization, rolling, urine marking, rubbing, and tail flagging. A temporary decrease in appetite may also accompany each heat period.

For owners with a cat in heat, patience may be tested to the limit. With all this "heat activity" comes the urge to stray from home and the owner to consider (with relief)—"good riddance." Since she is an induced ovulator, the female will roam until she is bred by the tom of her choice. Aside from the pregnancy, this wandering further exposes her to viral diseases and injury. This is also a significant contributor to the pet overpopulation problem—don't let it happen.

Grooming time for kitty. Even young members of the family can actively participate in caring for the family cat. Photo by Robert Pearcy.

A female cat does not need the experience of birth to be a happy, contented pet. Photo by Robert Pearcy.

The spay, or ovariohysterectomy, operation eliminates heat periods (and associated behavior changes) as well as the possibility of pregnancy. Solely interested in herself and her family, the spayed cat looks only for love, affection and care from the family. The spayed female requires less food, and her food intake and weight should be monitored to prevent weight gain from becoming a problem.

Too often, families obtain a female with the intention of having at least one litter, presumable to "make her better." Breeding doesn't affect her behavior, nor will it fulfill any emotional need. Pregnancy, to a cat, is just a physiological change that is internally regulated by hormonal checks and balances. As soon as the mother cat weans her kittens, her hormones quickly reset to "desiring" another litter. With this, she rapidly loses interest in the litter. Though the kittens will show a strong nursing instinct toward their mother for up to a year, she won't tolerate this behavior for long.

The major health concern that arises with the unspayed female cat is that she is particularly prone to intrauterine (womb) infections and breast cancer as she grows older. Not to be ignored, the unspayed female has an 85% chance of developing either disorder by 12 years of age. Spaying at an early age (preferably before the first heat) greatly reduces the chance of breast cancer, and spaying at any age eliminates the possibility of uterine infections.

The unneutered tom has an innate tendency to roam greater distances from home and spray urine to mark out his territorial perimeter. This marking acts as an announcement to all other males. It's usually satisfactory for him to spray trees and fencing when out of doors: but, if confined to the house, he will be quite comfortable using the drapes and furniture! Often invisible, the sprayed urine has a strong, easily recognized odor that is enhanced with humid air. Even when the tom consistently uses his litter box, the smell is pungent enough to permiate the home environment, and will linger even after careful cleansing of the box and replacement of the litter. It's rare for female cats to spray urine and mark territory, but it can occur. They use their litter box on a regular basis, or the garden when given outdoor privileges.

Castration, the surgical removal of both testes, will reduce the spraying in 90% of the cases and will eliminate the urine's odor within a couple of days. Of course, it also controls sexual desire, but experienced males may still show a passive interest in the opposite sex. Habits are easier to prevent than to resolve, so neutering by seven months of age is the common recommendation. Neutering also reduces roaming, fighting with

Manx kittens. In conformation, members of this breed are compact and well balanced. Photo by Robert Pearcy.

The unneutered tom has an innate tendency to roam greater distances from home. Photo by Robert Pearcy.

other males, and spraying. Approximately 10% of neutered males and 5% of spayed females will continue to spray, despite the surgery.

Male cats require particular attention to their diet to prevent feline lower urinary tract disease (FLUTD), previously termed feline urologic syndrome (FUS). FLUTD is a condition that results from many different causes that may be associated with the formation of "stones," or small, gritty crystals in the lower urinary tract. Although the exact cause of this disease has yet to be determined, the condition is best controlled by providing regular access to water and feeding a diet that is restricted in magnesium and promotes an acidic urine. Both male and female cats can develop FLUTD; however, male cats are susceptible to blockage due to anatomical differences. The urethra, a tubular duct leading from the bladder, in the male must pass through the penis. At this point, the urethra narrows substantially. The unloading of crystals into the urethra will cause the penis to become partially or completely plugged, creating an emergency medical situation.

SELECTING A BREED

Pedigree Cats

The purebred cat has a specific set of characteristics that set it aside from all other cats. When a male and female with these characteristics mate, they produce offspring with matching characteristics. This predictability of body shape, size, hair coat and colors, as well as behavioral traits, takes some of the guesswork out of choosing a pet for your home and lifestyle. For example, if you are away a great deal of the time then consider avoiding longhaired breeds, which require more grooming. On the other hand, a retiree may enjoy the sedentary

Facing Page: Male and female cats can make equally fine pets. Photo by Robert Pearcy.

Before you bring your cat home, you should have all of the accessories that you will need for its care. Photo by Robert Pearcy.

life of the longhaired Persian. Lifestyle may be more crucial than the breed of cat that you choose. Your living plans may require modification for the cat rather than the other way around.

Date of birth, ancestry and current veterinary care are well documented with pedigree animals and are absolutely essential to possess if you plan to show or breed your cat later. With a pedigree cat, you should ask about common genetic disorders found. Crossed eyes in the Siamese or deafness in all blue-eyed, white cats are examples. Interbreeding in purebred cats isn't uncommon to "improve the line." This unfortunate practice also increases the chances of getting the most undesirable inherited defects. Study the breed, become knowledgeable and be able to anticipate possible abnormalities before making this lifelong commitment.

While there are numerous advantages to choosing a pedigree, there are also some hitches. Availability is usually the prime difficulty. However, this varies from one breed to the next. Some, the Canadian Hairless, or Sphinx, in particular, are very new, having just been accepted by The International Cat Association (TICA) in 1987. Purchase costs vary greatly depending on the breeder's investment in the cat for sale. Popular breeds may be just a matter of searching the pets section of the classified ads,

Playing is hard work! Kittens will spend hours with each other—playing and wearing each other out. Photo by Robert Pearcy.

driving to the breeder's home and selecting a cat.

There is a natural bias on the part of some breeders about the desirable characteristics of the purebred cat being offered, and they may be overemphasized. "Good with children" and "of show quality" are the most common.

Many times, the conscientious breeder will keep the kittens until they are twelve weeks of age, primarily in order to select future quality kittens before being sold. This gives the kittens time for a good healthy start and allows time to get the vaccination program underway. However, this may also be a delaying tactic to buy time for treatment of parasitic, viral, bacterial or fungal diseases that are not easily detected. Some problems can even be endemic to a cattery.

Sources of purebred cats can be obtained from the following:

- American Cat Association
- American Cat Fanciers' Association
- Canadian Cat Association
- Canadian Cat Fanciers' Association
- Cat Fanciers' Association
- Cat Fanciers' Federation
- The International Cat Association

In the cat world, there are a few sales people on the fringes who are long on promises and short on paper work. However, most are reputable breeders who

Cats and very young children should always be supervised when they are together. Photo by Robert Pearcy.

A charming mixed breed kitten. In mixed breeds, you can find a variety of colors and hair coat lengths. Photo by Robert Pearcy.

stand behind their product and are extremely concerned about the cats and their future. Be prepared to take time and money to research a healthy animal and await its arrival.

Mixed Breeds

A mixed breed is defined as the progeny of two or more different breeds of cat. The common cat, generally referred to as "the domestic," with either a short, medium or longhaired coat, is the result of years of genetic outcrossing. With this huge gene pool, you can expect to find, within the same litter, a variety of colors and hair coat lengths. It's usually possible to find a selection conveniently located in any neighborhood. This is the pet that will be advertised as "free to a good home." The private home environment often keeps kittens isolated from contagious diseases and can provide a good healthy start. In any home, a litter is usually a novelty and all the handling they receive, especially in the first few months of life, helps to establish a normal, friendly relationship with people.

Generally speaking, inherited disorders are less likely in mixed

Patterns and markings can vary greatly in mixed breed kittens—even among members of the same litter. Photo by Robert Pearcy.

breeds. The potential problems with the mixed breed come primarily from the unknown ancestral background and frequent lack of veterinary care. Generalizations about behavioral traits are guesswork at best, since you rarely get a chance to observe both parents. You may be lucky enough to evaluate the kittens' mother. Expect some of the best characteristics with a mixed breed but be prepared to find some of the worst.

SOURCES OF CATS

Breeders

Experienced and established breeders give knowledgeable information about individual selection and immediate home care. Documentation of pedigree and current health status is available to guarantee the health and quality of the stock. The breeder's reputation is at stake with every pet sold, so one would expect higher standards for breeding and honest business practices.

Both parents are most often visible for a general idea of the mature appearance and temperament. Ancestral lines can be followed for future reference if breeding is planned. Registration papers, which you pay extra for, are necessary when you come to register your first litter.

Conscientious breeders are scrupulous about protecting their catteries from some of the troublesome diseases that appear months or even years after exposure. Feline Leukemia Virus and Feline Infectious Peritonitis are examples.

Unfortunately, amateur facilities may unwittingly harbor contagious animals and hence can pass the disease onto other pets. Fungal and upper respiratory diseases are classic examples of illnesses endemic to some catteries. Visit the breeder. Is the facility clean, bright and odor-free? Where possible, speak with previous purchasers to get an idea of the breeder's reputation.

A sale here is often final, and money is unrecoverable. You can't return an ill or undesirable pet unless you get a guarantee in writing. A veterinarian's evaluation, often within 48 hours of purchase, is required to validate any form of guarantee. Try to arrange for a trial period of one week.

Shelters and Pet Stores

Local and with convenient business hours, most shelters and pet stores have an abundant supply of cats along with all the supplies needed to set up your home for the new family member. Multiple kittens are on display—priced anywhere from free to a few dollars. This may even include primary immunizations and some cursory veterinary care.

"Spayed or not?" is often the question with adult cats. Evidence of surgical procedures may not be visible, and the "wait and see" plan can inconveniently bring you a female in heat or a surprise pregnancy. Specific arrangements can be made to have the procedure performed. Most shelters mandate the spay or neuter procedure at the appropriate age. Sexing kittens depends on the experience of the staff. You may find that "Sally" will become "Sammy" during the first visit to the veterinarian.

The nature of the shelter business is to accept strays and abandoned pets along with their medical problems. Contagious diseases are therefore prevalent. The animals also may not be very clean and hence require immediate grooming, insecticidal baths, ear treatment or deworming. On the plus side, shelters generally allow for a home-trial period. "All Sales Final" is the usual policy at the local pet store. Ask what their policy is.

HOW TO MAKE THE ALL-IMPORTANT FINAL SELECTION

Most people find that, in reality, their cat chose them, but if you do get the time to think about a choice then look for:

1. Clean, bright, shiny eyes without being runny or crusted with discharge.
2. No thick nasal secretions or sneezing.
3. Pink gums and white teeth.
4. Properly aligned upper and lower jaws.
5. Clean ears that have no odor, scabs or discharge. Avoid a kitten that consistently shakes its head or scratches its ears.
6. A hair coat that is clean and free of matting, dandruff and body odor.
7. An abdomen that appears full without distention. The anus should not be red or soiled with feces.
8. The body should not feel bony but yet be rather lean, firm and well fleshed over.
9. A history of a good appetite (without vomiting) and formed stools.
10. An interest in spontaneous, playful activity such as following a bow on a string or curiously watching your fingers scratching the floor.

This bright-eyed kitten radiates every outward appearance of good health. Photo by Robert Pearcy.

Abyssinian kitten. The large, wide-set eyes are characteristic of the breed. Photo by Robert Pearcy.

11. Your hand should be able to approach slowly from overhead without the cat crouching down or even folding its ears back. Avoid the kitten that growls or strikes out with its forepaw in a hostile manner.
12. Stroking of the head and back several times should be tolerated without turning to bite or scratch your hand.
13. Gentle handling should be permitted without aggressive biting or yowling.
14. Be sure it is able to not only run but also to walk with coordination and balance.
15. Be wary of shy, timid, fearful, or overly aggressive kittens.

Birman. Also known as the Sacred Cat of Burma. Accenting the silky coat is the contrasting color of the points. Photo by Robert Pearcy.

BREED PROFILES

Pedigree cat behavioral characteristics have never been clinically documented, and the following comments are based entirely on conventional wisdom. Consequently, you should expect to find some variations to these generally accepted norms. The breed standards are somewhat similar with each association, but they also have many differing requirements.

Abyssinian

These cats are of medium build, but well muscled. A gently rounded head with well-formed cheeks and rusty red nose over a whitish mustache and beard makes a particularly attractive face. Each hair has a characteristic ticking with a minimum of two bands of dark color. There were originally only two colors, ruddy orange-brown and rusty red. Today other colors such as blue, lilac, chocolate and silver have been developed. White spots on the neck or dark lines on the legs are not desirable, neither is a long pointy face. The Aby's durable coat requires little care short of stroking with your hand or a silken fabric and the occasional combing with a wire brush. "Curious" best describes their personality and gymnastic behavior. Gentle and good with children, they thrive on attention. Abyssinians make your home a playground—beware, they need lots of room to roam and investigate.

Birman

A relatively scarce cat that's worth waiting for, the Birman is the intelligent legendary temple cat. These cats have long silky body hair that seldom mats. Unlike many cats, this breed always enjoys the attention of being brushed and combed. Birmans' entertain themselves with simple toys and look forward to human companionship at any time. The four white paws, along with a face mask that outlines baby blue eyes, are the Birman marks of distinction. Seal, blue, chocolate and lilac are the recognized color points for the ears, legs and tail. Look for a round head, a well-developed chin and jaw and an almost Roman nose line. The body is stocky, long and usually considerably larger in the male. Birmans are quite tolerant of being handled by children but are quick to hide when the going gets tough.

Burmese

This golden-eyed, sable brown, shorthaired cat is now recognized in champagne, blue and platinum coat colors. Burmese cats have round heads and wide-set eyes, with a noticeable break over the nose to the forehead. With the exception of the blue Burmese's coarser hair, the satin-like coat sits tight to the muscular, athletic body.

Occasional grooming is all that is needed for this happy-go-lucky "monkey" of a pet. With its meek voice, this very affectionate breed demands attention that is hard to resist.

Cornish Rex

A perfect pet for the person who dislikes cat hair and has little time for grooming. Far from hairless, the Rex has a soft wavy coat made up of only undercoat hair—no guard hairs at all! The texture of the hair makes the Cornish Rex cat a delight to stroke. These cats enjoy being handled by persons of all ages. The muscular body, which feels cuddly and warm to touch, must have more temperate care in the colder climates.

The startling appearance of the Cornish Rex's oversized ears and seemingly small head appear almost disproportionate for this otherwise well turned out body. The Roman nose and slightly oval eyes suit this intelligent looking cat. Don't be surprised if the Rex shows explosive bursts of energy. Of course, that is when it isn't sleeping underneath a blanket.

Himalayan

This is another longhaired cat that enjoys a sedentary life. The Himalayan's round head, broad

Facing Page: A pair of Cornish Rex. These small to medium cats are distinguished by their soft wavy coats, which have no guard hairs.

face, prominent forehead, very short nose and smallish ears set low to the side make it quite Persian-like. A large body carries a full, fine coat, especially over the neck and shoulders, tending to make their front legs appear short. Like the Birman, Himmies have blue eyes, combined with all the colors of the Persian and the points of a Siamese. A cat of show quality will not have white hair anywhere.

Fortunately, this mild-mannered cat allows the essential daily grooming to be performed—some even enjoy being bathed! The Himalayan makes a most photogenic cat.

Korat

The Korat cat is a native of Thailand. The first pair of breeding animals were imported to the United States in 1959. The eyes are oversized and expressive, and eye color develops from kitten blue through amber, with green tinge to luminous green at maturity. The coat is flat, silky and close-lying and exists in one color: silver blue.

Korats make great pets, although they are not always easy to come by (due to their relative rarity). Many people who have allergies to other felines can tolerate this breed better. Korats do best on their own with a family, or with other Korats. They

Korat. This breed, which originated in Thailand, sports a beautiful silver-blue coat. Korats are pleasant-tempered and make fine companion cats. Photo by Robert Pearcy.

This trio of Himalayan kittens could steal any cat lover's heart away. Photo by Robert Pearcy.

Facing Page: Himalayan. This impressive-looking feline is essentially a Persian with the markings of a Siamese. In general, "Himmies" are quiet, easy-going cats. Photo by Isabelle Francais.

tend to dominate cats of other breeds in the house.

Maine Coon

This rough-and-tumble, larger than average cat has a broad chest and a long, firmly muscled body. The hair coat is very full and somewhat longer over the abdomen, hind legs and tail. The shorter hair over the neck and shoulder emphasizes the strength and ruggedness of the breed—one of the oldest in North America and probably our first working cat. The original "mouser," Maine Coons have a longer than average head but a noticeably square muzzle probably to better handle their prey.

Brown tabby is the usual color of the Maine Coon, but the variety is almost endless

Maine Coon kittens. Legends abound about this hardy longhaired cat that supposedly originated from Down East. Photo by Robert Pearcy.

although never chocolate, lavender or Siamese patterned. Its coat requires regular combing, especially in the areas of long fine hair. Maine Coons are comfortable anywhere from a boat to a barn and enjoy companionship with people and other pets. Their quiet voice adds an unexpected charm from this "one tough hombre."

Manx

One of the oldest registered cat breeds in the world, the Manx is genetically tailless. This characteristic is caused by a mutant dominant gene called, not surprisingly, the Manx gene. One fourth of Manx-to-Manx breedings do not survive birth because they inherited two Manx genes. This produces an anomaly of the anal area along with other vertebral malformations. Some Manx cats, called "stumpies" because of their short tails, are used for mating to assure survival of the breed.

A coat of many colors certainly applies here and most are accepted. The only exceptions are chocolate, lavender, and Siamese pattern, or these colors in combination with white. Manx have a distinct roundness to their overall shape. Emphasis is placed on prominent cheeks (fat whisker pads), a slight dip from the forehead to the nose, and round eyes. The rump is rounded over hind legs that are higher than the forelegs, which appear most stubby. Breeders look for a short

Maine Coon. The large eyes, high-set cheekbones, and lightly feathered ears are the distinguishing facial characteristics of Maine Coons. Photo by Robert Pearcy.

Manx. Taillessness is the distinguishing characteristic of this breed. Photo by Isabelle Francais.

coat made up of outer guard hairs and a fluffy undercoat. If Manx cats didn't have medium-sized ears with rounded tips, they could pass for rabbits in the woods.

Manx cats eat well and require only the simple care of a routine wire brushing. They make great apartment pets for a senior person, as they are quiet and have a tendency to buddy up to people.

Persian

This extremely popular longhaired cat is bred in six main color divisions: solid, shaded, smoke, tabby, parti-color and Himalayan points.

Odd-eyed white Manx. Photo by Isabelle Francais.

Each division has a wide range of specific colors.

The classic Persian has a round massive head with a wide skull and a flat forehead with a distinct break to the short nose. The smallish ears contrast with the large round eyes. A rather large deep-chested body with broad shoulders and a muscular rump is desirable.

A flowing coat needs daily wire brushing to prevent matting. It is imperative to comb out the hair until it is absolutely free of tangles before bathing.

Generally, Persians enjoy the quiet life and a full stomach. Relaxation and surveillance tend to be the high point of daily activities, yet they do have occasional playful moments without pestering for attention.

Persian. This elegant cat can add charm to any household. Persians are thoughtful, introspective cats that relish a tranquil environment. Photo by Robert Pearcy.

The Persian's timeless beauty and grace have made this breed one of the most popular in the cat fancy. Photo by Robert Pearcy.

The Siamese, slender, lithe, and long, is the embodiment of elegant conformation in a cat. The Siamese is intelligent and responsive and is said to be "the closest thing to a dog in a cat." Photo by Robert Pearcy.

Siamese

The color "points" of this breed are the feet, legs, ears and a face mask. Noticeable features include a wedge-shaped head with a straight forehead blending into the nose without a break, large pointed ears, blue eyes and a long slender body with forelegs slightly shorter than the hind legs. Siamese cats are recognized and registered in four coat colors:

- Seal—Either brown or black points and fawn body.
- Chocolate—Dark points and pale body.
- Blue—Deep silver or gray-blue points and a gray body.
- Lilac—Pinkish gray points and an almost white body.

An unusually intelligent, energetic, affectionate cat that's very vocal. Siamese can be dog-like and walk on a leash but really love a free, acrobatic lifestyle.

One unfortunate fault in the behavior of Siamese and Siamese-cross cats is their habit of "wool sucking." They are known to make large holes in socks, sweaters and blankets—especially dirty laundry.

Prone to dental gingivitis and halitosis, Siamese cats require conscientious maintenance of oral health.

SUMMARY

Choose your cat as you would a best friend. Keep in mind though, the seriousness required to safeguard its life and your commitment to being a responsible pet owner. Never surprise someone else with a kitten or cat as a gift. Inquire first whether this person is prepared to make the investment in time and effort to care for a pet.

The willingness to give attentive dedication to your cat's health and welfare will give it the chance to reward you with endless hours of devoted affection and years of loyal companionship. When you are ready, go ahead! Make this selection, but now with more knowledge and understanding.

ADDITIONAL READING

Clark, R.D. *Medical, Genetic and Behavioral Aspects of Purebred Cats*, Forum Publications Inc., St. Simon's Island, Georgia; Fairway, Kansas, 1992, 253pp.

The Allure of the Cat (TS-173) by Richard H. Gebhardt and John Bannon, over 350 color photos, 304 pp.

Atlas of Cats of the World (TS-127) by Dennis Kelsey-Wood, over 350 color photos, 384 pp.

The Mini Atlas of Cats (TS-152) by Andrew DePrisco and James B. Johnson, over 400 color photos, 448 pp.

Dr. Ian Dunbar is a veterinarian, animal behaviorist and dog trainer, with three books, fifteen booklets and eight videos to his credit. Dr. Dunbar received his veterinary degree and a Specials Honours degree in physiology and biochemistry from the Royal Veterinary College (London) and a doctorate in animal behavior from the Department of Psychology at the University of California, Berkeley, where he spent ten years researching the development of hierarchical social behavior and aggression in dogs. Presently, Dr. Dunbar writes the American Kennel Club Gazette's "Behavior" column, which was again voted best dog column by the Dog Writer's Association of America. Also, Ian writes for Dog Fancy, Cat Fancy, Dogs in Canada and for Dogs Today in England, where each summer he films his television series about dogs and people. Dr. Dunbar is currently the Director of the Center for Applied Animal Behavior in Berkeley, where he lives with a Malamute, two mutts and a kitty called Mitty.

How Cats Learn: The Role of Rewards and Punishment

by: Ian Dunbar PhD, BVetMed, MRCVS
Director, Center for Applied Animal Behavior
2140 Shattuck Avenue, #2406
Berkeley, California 94704

INTRODUCTION

Years ago, when needing some peace and quiet to finish writing a book on, dare I say it, dog behavior (Meow! Hiss-s-s!! Scr-r-r-atch!!!) I hid away in a friend's temporarily vacant apartment in Seattle. All I had to do was look after my friend's cat called Picasso. Picasso was a good-natured, well-behaved low-care cat. All he required was: breakfast and dinner -de nada; the window opened to let him out at 11pm -no problema; and the window reopened at midnight to let him back in -mi casa es su casa. Although I do not necessarily approve of cats unattended on public property, I had to respect the owner's wishes and the cat's routine and preferences and so, on the first evening I dutifully opened the window for Picasso at 11 pm. So far so good. But unfortunately, there was no midnight rendezvous with Picasso on the windowsill. Understandably perturbed at losing my friend's cat, I began calling "Picasso come" and received some strange looks from passers by on the street below, but none stranger than when I took to the streets and called for "Kitty, kitty, kitty!" employing the generic term for kitty we use in England. And did I really think Picasso was going to run up to a strange man on the streets past midnight. I don't think so! Consequently, I went back to the apartment, put out a bowl of food and went to bed leaving the window open. Sleepless in Seattle, I heard Picasso sneak in and after a few minutes, I crept out of bed to close the window. The next morning I set about teaching Picasso to come when called.

Rather than embarrassing myself further and advertising to one and all I had lost the cat again, I decided to train Picasso to a 'silent' dog (Hiss-s-s-s!!) whistle. Peep-treat. Peep-treat. Peep-treat—all day long. I'll tell you that cat had it down quicker than any dog I've ever trained: Brain like a whip, recall like a whippet. And later at midnight, I opened the window and "Peep! Peep Pee-e-e-e-p!" Bothered no one. Embarrassed—not. And that cat flew inside to roost like a homing pigeon. Ah! Applied science—there's nothing like it!

WHY TRAIN?

There are many reasons why cat training (communication and education) is imperative. Cats should be trained for their own safety. A cat which has been trained to come when called is much easier to catch if it escapes from the house, car or veterinary clinic. It can be exasperating and hazardous to try to catch an indoor cat once it has escaped outdoors. Most owners panic and run after the cat, only causing the cat to panic and run away from the owner. A cat trained to meow on command is much easier to find if it becomes lost. Some cats may remain in hiding for days, or weeks at a time. Often they are very close to home and the owners have passed very close to the cat's hiding place. Scared cats may not come out of hiding when called, but they will often meow on command, if trained to do so.

Basically though, training is the means by which we teach the cat the meaning of the words we use. Obviously, it would be inhumane not to train. Imagine having to live in a foreign country without understand a word of the language. People might remark that you're very independent and aloof, almost quite snooty, "Well, he never goes out and never

Learning is a lifelong pursuit for most felines. Photo by Dr. Ian Dunbar.

Cats love nooks and crannies...any niche that they can call their own. Photo by Dr. Ian Dunbar.

socializes with people." It can be hard to socialize if you don't know the language. And so we owe it to our cats to teach them our language and to try to understand theirs. Contrary to popular belief, the cat is an highly social animal and it is just downright unfair to invite the cat to come and live with us but then to subject it to social isolation and solitary confinement.

The greater number of words from the human language cats understand, the easier it is for owners to communicate their wishes. For example, a cat which has been trained to go to its litter box, or to go to its scratching post, will be much less likely to soil the house and shred leather upholstery. Also, a cat, which enjoys regular vigorous training sessions at ten in the evening, will be much less likely to go stir-crazy at two or three in the morning. It would be unfair not to teach the cat our language and to keep house rules a secret, only to later punish the cat for breaking rules it did not even know existed.

You know, some misguided folk maintain, cats can't be trained. Oh no? Why not try explaining that to a lion tamer, when he has his head in the lion's mouth. If big cats can be trained, surely our little domestic kitty should present no problems. The major difficulty appears to be that cats do not respond to antiquated, push-pull, squish-squash dog training methods. Cats will not put up with this abusive rubbish.

Like all animals (with the exception of dogs and people), cats will just not tolerate an abusive relationship. In fact, we are very lucky domestic cats are not the same size as domestic dogs. If we had cats the same size as Malamutes and Mastiffs, there would be very few unscarred cat owners. Socialized dogs have a strong urge to apologize when they err. Thus reprimands can have a perverse but nonetheless effective bonding effect. Of course, if the dog is not socialized, then it will respond to psychological and physical abuse the same way cats do—by abruptly terminating both the training session and their relationship with the 'trainer'. It is silly to say cats are untrainable just because they dislike abuse. Surely it is the trainer which is faulty, not the trainee. Indeed, training a cat is a good test to see how much you know about animal learning. I have always said if you can teach a cat or a chicken to heel, you will be a better dog trainer and a better parent.

But it is still easier for some owners to make excuses, than to give it a go and try. Of course, this usually means, the owner hasn't got a clue how to do it. Well, it's time to learn how folks, because if you do not know how to train a cat, then you do not know how to train. Cats are really no different to teach than any other animal—they are immensely smart critters and take to intelligent education like, dare I say it... a dog to a bone.

Cats are very intelligent and can learn new things, depending upon how successfully they have been trained. Photo by Dr. Ian Dunbar.

A LITTLE TRAINING THEORY

Like all animals, cats learn by association and anticipation. When two events occur consecutively on a number of occasions, the cat begins to associate the two and begins to anticipate that one leads to the other, i.e., the cat learns, specific behaviors may be routinely associated with specific consequences (either pleasant or unpleasant). Moreover, the future likelihood or incidence of each behavior is influenced by its consequences, such that: If the outcome is pleasant, the cat will likely repeat or continue the activity, whereas if the outcome is unpleasant, the cat will discontinue or avoid the activity in the future. For example, if a cat is rewarded with a food treat each time it runs to its owner, the cat quickly associates approaching the owner with oodles of affection and food treats and consequently will become more sociable, seeking the owner's proximity in the future. If on the other hand, the cat was lying peacefully on the couch while the owner approaches malevolently screaming like a banshee and grabbed the cat by the scuff to rub its nose in a pile of cat doodoo, the cat will no doubt associate the owner's approach with idiosyncratic and extreme abuse, and will be unlikely to relax on the couch in the future. In fact it is unlikely the cat will remain in the same room as the owner. Perhaps the cat will opt for

Cats love to explore and investigate. Photo by Dr. Ian Dunbar.

an independent and aloof lifestyle, basically because they do not care for their owners. There is simply no loving relationship and so the cat decides to adhere to its own agenda. Insisting cats are independent, aloof, asocial, or even antisocial are common excuses for the owner's failure to provide the cat with companionship, compassion and caring instruction. How would you feel if you went to chat to a friend, only to have him grab you by the scruff and rub you nose in a pile of feces. Surely, that would hardly augur well for an enduring relationship. No wonder cats appear aloof and hide in closets. They are just fed up with human foibles and aggressive idiosyncrasies.

Additionally, cats learn to *associate* the consequences of their actions with specific stimuli or situations. If the owner punishes the cat each time the cat sharpens its claws on the back of the leather couch, the cat will likely cease and desist scratching the couch *when the owner is around.* But... when the owner is in another room, or away from the house, the cat will delight in honing its claws to stiletto points. This has little if anything to do with misbehavior, stubbornness, spite, vindictiveness, or *separation anxiety.* In fact, a more descriptive term would be separation fun—rather than learning its behavior is inappropriate in the eyes of the owner, the cat has simply learned to *associate* the unpleasant consequences of its actions (punishment) with the owner's presence, i.e., the cat is punished when the owner is present, but is not punished for the same activity when the owner is absent.

When training a cat, a verbal request (command, or instruction) provides the necessary stimulus to signal the cat, appropriate actions will likely be associated with rewarding consequences. Thus the training sequence comprises: **Request—Response—Reward.** For example, the owner requests the cat to scratch its scratching post (Request), and after the cat dutifully scratches the post (Response), the owner praises kitty and offers a food treat (Reward). Thus the cat *anticipates* rewards will be forthcoming when it responds appropriately following the owner's request. The only question remains, how on earth do you get the cat to scratch on cue, or to do anything on command for that matter? Simple! By waggling a lure high enough up the scratching post, so the cat has to stand on its hind legs to get its forequarters high enough to swat at the lure with its forepaws. After a few trials, the cat learns to *anticipate* that the lure-movement will follow the Request and consequently, it will run to its scratching post when requested to do so.

LURES AND REWARDS

Crinkly paper, a tassel, or a food treat in the hand, or at the end of a wand or fishing pole and line are the best lures to entice the cat to respond quickly and appropriately. Offering the lure for the cat to play with and especially giving food treats are the most effective rewards for novice owners to use in cat training. However, whereas many animals, including horses and dogs, readily accept food from the hand, most cats have to be trained to do so. Training a cat to accept food treats dramatically facilitates training. Practice by hand-feeding the cat's dinner. Put the first few bits of kibble one at a time in the cat's bowl to eat. Then place pieces of kibble on the floor. When the cat eats readily, place a single piece on the floor but rest your

Cats usually get along fine with dogs (above)...once they have taught them a few lessons (below). Photos by Dr. Ian Dunbar.

fingertip on top of the kibble. Now place the back of your hand on the floor with a single piece of kibble on the palm. Repeat this holding your hand a little higher each time and eventually, hold the kibble between finger and thumb.

A treat is not something extra, and it should not be junk food. Normal kibbled dry food (part of its daily diet) is fine. Weigh out the cat's food in the morning and place it in a plastic container with a little freeze-dried liver powder sprinkled on top to add aroma. Different family members may take kibble from the container to train the cat, with the knowledge the cat is receiving no more food than is normal.

ROLE OF REWARDS AND PUNISHMENT

Lure/reward training is the *sine qua non* of cat training. Punishments have absolutely no place in cat training. Punish the cat and you advertise your incompetence as a trainer and will likely lose the cat's respect and companionship. Perhaps you would be better off with a pet rock, or a cabbage as a companion. If the cat makes a mistake and gets it wrong, teach it what is right. It is easier, more efficient, more effective and certainly more enjoyable to teach and reward a cat for what is right, than to try and punish it for making mistakes and breaking rules it did not even know existed. Remember, there is only ONE RIGHT WAY, therefore it takes very little time to teach the cat what is right, whereas it would take an infinite amount of time to try to punish a cat for the infinite number of ways it can get it wrong. But even if you do punish the cat a few times, you are only teaching (forcing) the cat to resort to misbehaving (acting like a cat) in your absence. Additionally, unless you are a little-brained weirdo, administering punishments is not much fun. And cats don't like being punished either. Cats want to trust and enjoy their human companions.

COME WHEN CALLED

First decide on a suitable signal. Verbal commands, whistling and hand-clapping are ideal, because you can give the signal at any time. "Picasso, come here" is fine. Initial training should begin at dinner time. Take a handful of kibble and give one to the cat for free, to let it know the game is afoot. Hold out a second treat, say "Picasso, Come" and back away. The cat will approach to claim its reward. Repeat this several times, backing up farther with each repetition. If the cat does not come, take a break and postpone the training session until the cat is hungrier. Once the cat has the idea, you need not reward it each time. In fact the cat will respond more enthusiastically if it is only rewarded once every three or four trials. As a rule of thumb: If the cat always comes, reward it less often; whereas if the cat loses interest, reward it more frequently. Cats normally approach friends at a nifty pace during pleasant encounters but by reserving occasional rewards for the cat's speedier responses, the cat will approach faster and faster. And of course, turbo-recalls deserve bundles of affection, several food treats and maybe a special piece of freeze-dried liver.

Call the cat to you for all good times, such as dinner or an evening on the couch. Make a point of calling the cat to you whenever the telephone rings, when the kettle boils and always before opening the door to let it outside. This cat will be worth its weight in gold: You'll never miss another telephone call, you'll never let another kettle boil dry and if (heaven forbid) your house ever catches fire, even the most un-altruistic of cats will come to wake you to open the door so it can escape the blazing inferno.

GO TO... PEOPLE

Once the cat comes readily when called, it is time to make the game even more enjoyable. Two owners can go to opposites ends of the house and call the cat back and forth. Several owners may participate (perhaps some upstairs and some downstairs) to make the exercise a round-robin

A cardboard box can make a fine temporary resting place. Cats will amuse themselves with any number of common household items. Photo by Dr. Ian Dunbar.

recall. Only one person should give instructions to the cat at a time. At the start of the exercise, one person should call the cat and then indicate the next person to call by saying, for example, "Picasso, Go to Allen." Then it's Allen's turn to call the cat, to offer a deserved treat and then choose the next person to call and so on. After a few trials, the cat learns following a "Go to..." command someone is likely to call it over for a treat and so it may anticipate the recall and run to a person—any person—before it is called. Should the cat run to the wrong person, simply ignore it and say nothing. Let the right person continue to call and offer a treat when kitty arrives. Eventually, the cat will associate the instruction "Go to Kristie" specifically with Kristie's recall request. When the cat begins to anticipate the *specific* recall request, it is beginning to discriminate between different "Go to..." commands. Now if Allen is lounging in bed on Sunday morning, instead of making coffee and buying the papers, Kristie can tell kitty "Go to Allen" and in anticipation of a yummy food treat, kitty will wake Allen with a mid-chest impact on par with a Scud missile.

GO TO... PLACES

"Go to..." commands are extremely easy to teach and they have a myriad of applications for improving owner/cat communication. The cat may be taught to go to different rooms, to go inside, to go to different items of furniture, or to special kitty-care paraphernalia, e.g., litter box and scratching post.

There are a variety of techniques to teach a cat to go to its special chair, bed, cozy cuddler or traveling crate. For example, say "Kitty, go to your chair" and then use a food treat to lure the cat to its chair and pet the cat when it settles down *in situ.* Or, first place a treat in the traveling cage as a reward, give the command and then, entice the cat to its crate, where it finds the treat once inside. Two people may train the cat in much the same way: One person issues the instructions to the cat and the other person rewards the cat once it goes to the right place.

Teaching the cat to go to its litterbox and/or scratching post greatly facilitates the prevention and treatment of two of the most common behavior problems. Many of the "Go to..." commands eventually become activity instructions. For example, when training the cat to go to its litterbox, at first the cat only has to go to its box for a reward but later in training, the command will mean: Go to your litter box and use it. Similarly, during initial training, "Go to your scratching post" simply means: Go to the base of the post to claim a goodie. Later, the cat will be

Facing Page: Cats can do amazing things. This cat has learned to jump through the hoop on command. Photo by Richard R. Hewett.

required to climb on the post and scratch to get a reward. Eventually, scratching the post becomes a reward in itself.

PUNISHMENT

As mentioned, physical punishment has no place in cat training. No place whatsoever. Physical punishments quickly and effectively teach the cat to fear the owner. Also, they teach the cat only to misbehave when the owner is absent. So now the owner has an even more annoying and hard-to-correct problem. Instructive reprimands are sufficient for kitty misdemeanors, including, "Litter Box!", "Scratching Post!", "Off!", or "Outside". The tone of the reprimand informs the cat that it is transgressing but the instruction also reminds forgetful kitty what to do to get it right, provided that you have previously taught the cat the meaning of the instructive reprimand. Even so, the necessity of a single reprimand is advertisement; you have not sufficiently trained the cat in the first place. Bad owner! No chocolates!

By far the best tactic is to prevent problems by providing appropriate outlets for play, scratching, eliminating, sleeping and feeding and to teach the cat how to use them. A scratching post, perhaps some kitty greens to munch on, some proper cat play toys, or even a few empty cardboard boxes to play in should keep most cats occupied and out of mischief when the owner is away. Just put a motorized ball in a paper bag and the cat will be amused for hours on end.

During the training period, a few commonsense temporary precautions such as closing the odd door, rearranging furniture and using some child-proof (and hopefully kitty-proof) locks effectively keep the cat away from some of the more tempting items, such as plants, drapes or leather furniture. Failing that, a few strategically placed booby traps will convince the cat to stay away from specific areas. A number of commercial booby traps will be discussed later in the book. However, in most cases, all it takes is a little ingenuity. A few strips of double-sided sticky tape or a stack of empty soda cans set to topple are usually sufficient to keep a cat from jumping onto counters. Similarly, the bottom of the drapes, or a piece of netting placed over expensive furniture, may be tied to a few bottom cans in the stack. Booby traps are effective because the cat learns from the environment; the unpleasant consequences of the cat's bad behavior are not associated with the owner's presence and so the cat does not seek to misbehave in the owner's absence. And, the cat does not become afraid of the owner.

STAY IN THE YARD

Many outdoor cats become lost after they're moved to a new home. The first time they go outdoors, they often panic and run and hide. The cat may remain hidden for a number of days because it is afraid to venture into the open. Usually the cat is close to its new home but because it panicked when it ran, it may not know the way back. Also, if the owner did not allow the cat sufficient time to become attached to its new home, it may feel little need to return. Instead, it will most likely try to return to its former home.

After moving an outdoor cat to a new home, keep it inside for several weeks. When letting the cat outside for the first time, make sure it is hungry by withholding food for a day. Let the cat poke its nose outside for just a minute and then call it back inside for a tasty morsel and then let it outside once more. Repeat this procedure a number of times, letting the cat remain outside a little longer and to explore a little farther each time. Thus, the cat begins to develop the notion of staying close to home. In addition, the 'escape route' indoors becomes second nature to the cat.

Outdoor cats are exposed to many more dangers if they are allowed to leave your property. A variation on the above exercises may be used to teach the cat to respect the limits of its property. Weigh the cat's daily ration of food (canned and/or dry food) and divide it into twenty servings. Put one serving in the cat's dish, show it to the cat and put the cat outside. After just a few minutes,

Catnap. Like many other household pets, cats enjoy a good snooze in comfortable quarters. Photo by Dr. Ian Dunbar.

Catnip has long been a favorite of cats. Pet shops carry it in a number of different varieties. Photo courtesy of Cosmic Pet Products.

call the cat (or give a blast on an ultrasonic whistle) and if it comes within 30 seconds it gets to eat the first course. If it doesn't arrive within the allotted time span, show the cat what it just missed and put the first course back in the fridge, (as part of the next day's rations). Put the second course in the cat's dish, show it to the cat and put the cat outside again throughout the day. Repeat the procedure with each of the remaining food installments. Each time the cat comes in time, it gets to eat; each time it is late it forfeits five percent of its daily rations. Within just a few days, the cat will remain close to home. Basically if you want a cat to stick around, you've got to give it a good reason to stick around.

MEOW AND SHUSH

Most cats meow for attention and affection and whenever they want food. They meow when they want to go outside and they meow when they want to come inside again. To train your cat to meow on cue, predict when the cat will meow and then request it to do so beforehand and reward it immediately afterwards. For example, show the cat a tasty morsel, say "Kitty, Speak" and then offer the treat as soon as it meows. Alternatively, put the cat outside, or in a closet, ask it to meow and then rattle its food dish, or manipulate the can opener. The instant the cat meows, open the door and serve dinner.

When training the cat to meow on command, be careful not to create a monster. Never reward the cat for meowing unless you have requested it to do so first. Also, right from the start, teach the cat to shush at the same time as teaching it to meow. Teaching shush is a boon when living with any nocturnal critter. Reward the cat the instant it meows on the first few trials but thereafter, praise the cat for meowing and then request it to "Shush" and wait for one second of silence before offering the treat. On the next trial wait for two seconds of silence after meowing, and then three seconds on the third trial

Facing Page: Sweet dreams. Some cats are real "cuddlers" and love to curl up with members of their human family. Photo by Dr. Ian Dunbar.

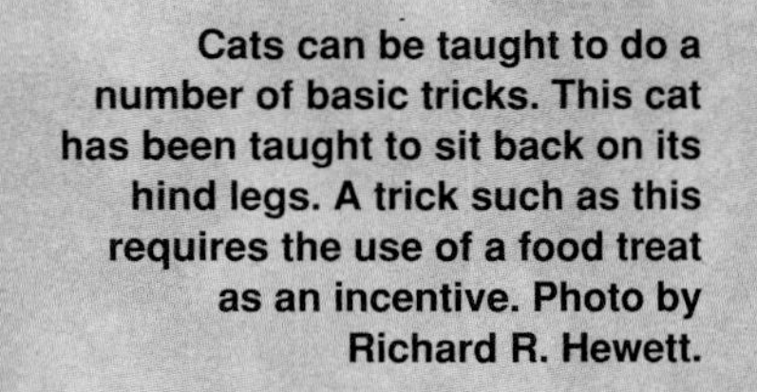

Cats can be taught to do a number of basic tricks. This cat has been taught to sit back on its hind legs. A trick such as this requires the use of a food treat as an incentive. Photo by Richard R. Hewett.

and so on. Basically, you are training a 'meow-shush sequence' and once the cat understands both commands it will be possible to alternate the two and have meaningful conversations with your cat. Also, the cat now at least understands what you mean when you ask (scream) for it to shush (shut up) in the middle of the night.

SIT, DOWN, STAND AND ROLLOVER

Teaching a cat to assume different body positions on cue is easy, enjoyable and essential: Cats enjoy the game, family and friends love watching the cat have a good time, and training the cat to stand still and rollover and lie on its back are essential for examinations in the veterinary clinic. A veterinarian can not perform his/her job properly unless the cat is at least calm and tractable. It is unfair to both the cat and to the veterinarian not to prepare the cat for veterinary examinations.

Start with the cat standing, Say "Kitty, Sit", waggle a food treat in front of the cat's nose and then move the treat upwards and backwards over the cat's head. As the cat's nose follows the treat, its neck will bend back, causing the cat to sit. Praise kitty and offer the treat as soon as it does so. If the cat rears up on its hind legs and swats at the treat, you are holding it too high. Lower the treat until the front paws touch the ground. (Hold the treat higher to teach the cat to rear up and dance and give high fives to oust Benji for Hollywood auditions some other time.) Training the cat to lie down is best done on a table. Lower the treat just below the table edge out of reach from its jaws and paws and the cat will lay down so that it may reach down further with its forepaw. Move the treat backwards from the cat's muzzle to its withers, and as the cat looks backwards over its shoulder, tickle the cat's abdomen and it will rollover for a tummy-rub. Pull the treat away from the cat and hold it six to eight inches above the table top and the cat will resume standing. Repeat the four body positions in random order, progressively increasing the length of time the cat has to remain in each position to claim its treat. It's as easy as that.

ADDITIONAL READING

Bohnenkamp, G: *From the Cat's Point of View*, Perfect Paws, San Francisco, CA, 1991, 48pp. Available from James & Kenneth Publishers, 2140 Shattuck Avenue #2406, Berkeley, CA 94704.

Bohnenkamp, G: *Kitty Kassettes* (Audiotapes). James & Kenneth Publishers, 2140 Shattuck Avenue #2406, Berkeley, CA 94704. 25 min x 5.

Dunbar, I; Bohnenkamp, G: *Cat Training*, James & Kenneth Publishers, 2140 Shattuck Avenue #2406, Berkeley, CA 94704.

Dr. Clayton MacKay is a graduate of the Ontario Veterinary College in Guelph, Ontario and now serves there as the director of the veterinary teaching hospital. In addition, he is the President Elect for the Board of Directors for the American Animal Hospital Association and the hospital co-director for MacKay Animal Clinic. Dr. MacKay has made hundreds of presentations to audiences throughout Canada and the United States and is the current President of the American Animal Hospital Association.

Preventing Kitten Problems

By Clayton MacKay, DVM
Director, Veterinary Teaching Hospital
Ontario Veterinary College
University of Guelph
Guelph, Ontario, Canada N1G 2W1

INTRODUCTION

Wow! So you've decided to get a new kitten. I'm sure all of us give it some thought every time one of those fantastic commercials featuring several gorgeous kittens plays on TV. Unfortunately, for every kitten given a home, there is an equal number of cats found stray or turned over to adoption agencies and shelters because the owners were unprepared or unwilling to follow through on the lifetime of care. Taking any living creature into our homes should be a decision based on more than just the emotion of the moment. A little planning can make this new creature a pleasure as a kitten and set up a lifetime of enjoyment as an adult cat.

Before bringing your new kitten or kittens home, be sure to have followed the advice on how to choose and select a healthy one. If possible, always have your new kitten examined by your veterinarian prior to it entering your home. This can keep you from starting with a health problem and avoid external parasites being brought into your home with your new friend.

Although I have found little difference in the behavior of neutered or spayed cats, there are some people who insist on having only males or females. Early in life, it is difficult to be sure which sex your kitten is, so please have it checked by your veterinarian. Tri-colored kittens are almost always female; red or orange are male in

A cat carrier is a necessity for the cat owner. Cat carriers are indispensable when it is time for vet trips, vacations, and the like. Photo by Isabelle Francais.

Cats are agile, acrobatic, and lightning quick. Their reflexes enable them to easily leap at—and catch—moving objects. Photo by George Pickow from Three Lions.

You can choose from a variety of cat litters. Whichever brand you choose, be sure to change it on a regular basis. Photo courtesy of Hagen.

about 70 percent of cases. Totally white kittens with blue eyes can be congenitally deaf.

PREPARING FOR YOUR KITTEN

Be sure to do your homework before you actually pick up your new kitten. Pick up cat litter and trays along with appropriate foodstuffs. If you can find out what the kitten is being fed, try and choose the same product and then slowly change to the product you would like to use. This gradual changing of diet can often avoid stomach upsets or diarrhea.

Find a place in your home for the kitty litter tray or trays. It should be in a quiet and traffic-free area if possible. Too much noise and traffic may discourage the kitten from using its new bathroom. If this early litter training is not done, it will be a frustrating exercise of trying to retrain. Litter should be odor and dust free. There are several clumping litters that allow you to scoop both bowel movements and

There are a number of products available that will help to keep your cat's litterbox fresh smelling. Photo courtesy of Hagen.

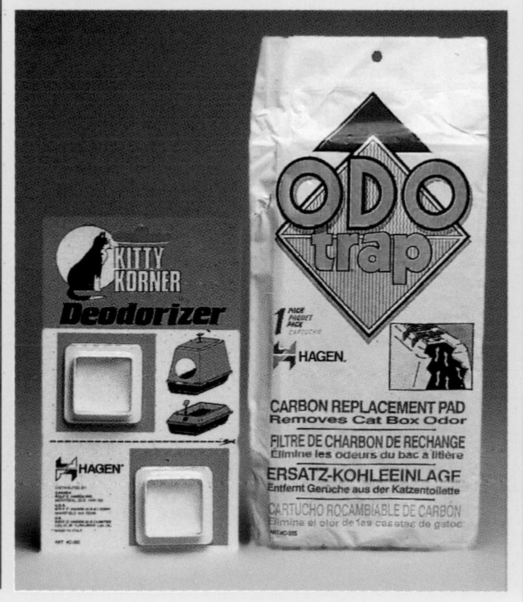

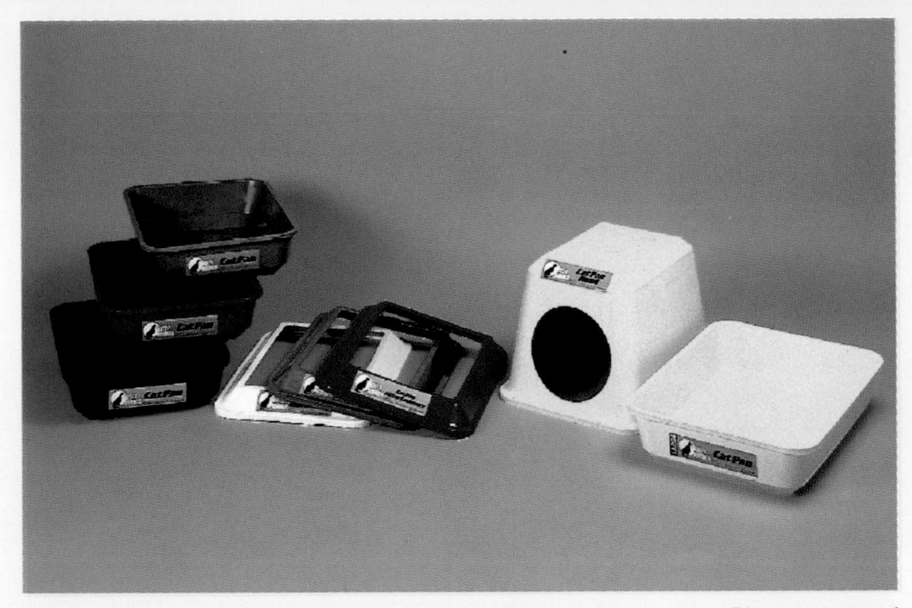

When selecting a litterbox, look for one that is durable and easy to clean. Photo courtesy of Hagen.

urine. Some cats will not use litter is there if any waste material already in it. Therefore the scoop-and-clean type allows you to keep the box clean.

Take assessment of your home with the thought of having your new, curious, and playful friend. For at least the first six months, have fragile knickknacks and family heirlooms hidden away. Scout the home, looking for areas that could be health hazards if encountered by the curious kitten. Watch out for open motors, sump pumps, electrical cords and areas where a tiny cat could crawl in and be trapped. Don't forget to check the car, the washing machine and the dryer. Check doors and windows to prevent escapes; and if you live in an apartment in a high rise, be sure to secure the balcony area. Although there are some miraculous stories of cats surviving high-rise falls, the majority are either badly injured or killed.

Now that you have checked out all of the above, it is time to pick up your new kitten. Be sure to have acquired a cat carrier. If you start by using this device on your very first traveling experience, it often means you will be able to take the cat with you from then on. Often, cats are not good travelers because of an early bad experience. Today we are even allowed to take cats in the cabin of many airlines, so chose a carrier that will fit under an airline seat. Place a small blanket or towel in the carrier for the kitten to snuggle into along with a few pieces of dry food for a treat.

In some cases, it is wise to have a kitten/cat proof room. This should be a small area where the cat can be isolated for a period of time and unable to harm itself or the surroundings. Food and kitty litter, along with a few suitable toys, would be the only things in this room. It is often wise to use a small room like the second bathroom or utility room. There should be no electrical cords or breakable items and hopefully no draperies or other finishings that could be damaged. Child-proof locks can be used to close off cupboards. This area is especially useful when you are unable to supervise your new housemate and to praise or scold behavior as it occurs. Animals have no language

In some situations, you may have to kitten/cat proof a room. If the room has cabinets, you can have them fitted with child-proof latches to keep kitty out. Photo by Dr. Gary Landsberg.

A pair of Abyssinians ready to travel. Photo by Isabelle Francais.

skills, and therefore you cannot have meetings with them. Praise of good behaviors and scolding bad behaviors must be delivered at the time of the behavior to be effective.

For those who have lifestyles that require that they be away from home for long periods of time, I would suggest having two kittens. This companion is then the prime source of entertainment. It can sure save on the wear and tear on the household environment, and it has been my experience that there is very little added work or expense associated with a second kitten.

Mother cat and her kittens at play in the backyard. Interaction with its mother and littermates is essential to a kitten's development.

UNDERSTANDING KITTEN PLAY

Many of the actions of the kitten in play that are so much fun to interact with are instinctive and designed to allow the cat to survive in the wild. The chase-and-catch play mimics the actions they will need to feed themselves if necessary. Cats seem to have maintained their wild instincts to a greater degree than our other favorite pet, the dog.

Like the young of other species, the young kitten seems to be either playing flat out or sleeping. They go through bursts of wild energy followed by almost comatose sleep. Again, like most other young, the amount of sleep required decreases as they grow. Cats are, by nature, nocturnal (preference for nighttime activity). It is therefore important to try and keep the kitten busy and active so that it will establish a daytime routine and be more inclined to sleep at night.

As soon as you get the kitten home, gently clip the tiny nails. It would be wise to have your veterinarian show you how this should be done to avoid clipping too close. What you are doing is preventing the new kitten from attaching itself to various parts of your body and house by these tiny needle-like claws. The hiding and pouncing, chasing and grabbing,

Cat play area with toys. The top cat is on his favorite perch. Photo by Denise Barron.

This cat, Luther, and dog, Grace, have been raised together since they were young and are best buddies. Photo by Dr. Gary Landsberg.

jumping and latching on are all behaviors programmed in as survival patterns. While you may and should discourage some of this, you must realize that normal behaviors are almost impossible to stop. The goal therefore is to direct the play into appropriate channels.

You will find many kittens and cats that exhibit a "kneading" behavior with their front feet. This inherited trait functions to remove extra nail tissue and deposit scent from the sweat glands in their feet. This behavior is often done with the cat standing on its hind feet and moving the front feet in a raking motion. Because these are normal traits, "training them away" is almost impossible. If you are unable to get your cat to use a scratching post, then you should discuss other options with your veterinarian.

USE OF TOYS

Having an armament of toys for your new friend will help use up the boundless energy in a non-destructive manner. Set up small objects on strings tied to door handles and drawer openers. The kitten will swat away at them for hours while really playing by itself. Small Nerf®-type balls can be batted and chased. There are even small toy fishing rods that enable you to cast out a safe "bait" for the kitten to either capture in mid-air or pounce on as you troll it slowly by the kitten's hiding place. All of these

You can choose from a wide variety of good toys for your kitten, including various types of balls and "fishing poles" with amusing toys attached to the end of the line.

toys reinforce the fact that the kitten is not attacking its owner—hopefully. The intention is to direct the play toward acceptable toys and hopefully away from the owner's feet and hands. By being proactive with setting up the play behavior, you will be able to continue these games for life. Oftentimes when there is inappropriate behavior, you can simply have the kitten change to a more acceptable game.

EXPOSE THE KITTEN TO MANY NEW THINGS

Try and expose your kitten to as many new people, objects and noises as possible in its formative months. Generally, by allowing the kitten to experience various people and situations at an early age, it will be more adaptable with age.

This will give you more freedom to go away for a weekend or a long trip and have someone else care for your pet without worry of strange behavior or stress to the cat.

SUMMARY

Raising a cat is much like raising a child and takes many of the same efforts. This chapter reviews some of the more important ways you can prevent problem behaviors from developing in your kitten. Remember, it is always much easier to prevent a problem behavior than to correct it afterward.

ADDITIONAL READING

Beaver, B.V.: *Feline Behavior: A Guide for Veterinarians.* W. B. Saunders, Philadelphia, 1992.

Eckstein, W.: *How to Get Your Cat to Do What You Want.* Villard Books, New York, 1990, 263pp.

Jester, T.: *Train Your Cat.* Avon Books, New York, 1992, 110pp.

The Allure of the Cat (TS-173) by Richard H. Gebhardt and John Bannon, over 350 color photos, 304 pp.

Atlas of Cats of the World (TS-127) by Dennis Kelsey-Wood, over 350 color photos, 384 pp.

The Mini Atlas of Cats (TS-152) by Andrew DePrisco and James B. Johnson, over 400 color photos, 448 pp.

Dr. Patrick Melese is director of Veterinary Behavior Consultants, a veterinary practice dedicated to preventing and treating behavioral problems in animals, and the Tierrasanta Veterinary Hospital, a general practice in San Diego, California. Dr. Melese's training includes both Bachelor's (1979) and Master's degrees (1980) from the Department of Zoology at the University of California, Davis where his subjects were Animal Behavior and Neuroscience. From 1980 to 1982 Dr. Melese worked as a research associate, doing behavioral research and publishing papers in physiological psychology. Dr. Melese continued to participate in behavioral research and assisting with the Behavior Service in the Veterinary Teaching Hospital while he attended veterinary school. Dr. Melese earned his Doctorate in Veterinary Medicine from U.C. Davis in 1986.

After receiving his veterinary degree, Dr. Melese has been a veterinary medical, surgical and behavioral practitioner. He has kept abreast of advances in behavior literature, taught classes in veterinary behavior, given presentations to public and professional groups, and consulted with colleagues in the field throughout the world. For the past several years he has developed and taught a 6-week behavior class for young puppies (age 10-18 weeks) which is currently taught in several locations in San Diego County.

Kitten Problems and Kitten Training

By Patrick Melese, DVM, MA and Carol M. Harris
Veterinary Behavior Consultants
10799 Tierrasanta Blvd.
San Diego, CA 92124

INTRODUCTION

Getting a new kitten can be a fun experience and at the same time a trying and frustrating one. Much of this depends on the pet that you select and how you plan for success. For example, if you are a low-key, relaxed person, a highly energetic over-exuberant kitten may be inappropriate for you. You may be inviting behavior that will drive you crazy. On the other hand, if you are a high-energy person, a couch potato kitten may disappoint you.

BASIC TRAINING

The main factor in avoiding play-related problems is to provide plenty of outlets for your new kitten to act in ways that work for your family and your baby feline. Never encourage behavior (such as aggressive play with hands and feet) that may lead to future problems. If you wish to avoid a life of yelling ("Here Kitty, Kitty, Kitty..."), hoping your cat will come to you, take a few moments and teach your kitten to listen to you and to respond to its name and a couple of simple commands. This establishes people as social beings who should be listened to and who have wants that should be considered. Always use the kitten's name prior to a pleasant event such as play or feeding. It can take the form of a command like "Tigger, Come Here," followed by preparing the kitten's food, or reaching for a toy to play with your pet. Even a brief, friendly stroking session of rubbing your pet's favorite spot can be the reward for coming if your cat likes such contact. Never follow a request for your kitten's approach ("Come Here") with anything your little friend dislikes (for example, trying to catch him by chasing) no matter how important you feel it is, or the kitten will learn to avoid you following the "Come Here" request. If the kitten does not come promptly, encourage it to respond quickly by getting up and showing your little cat how wonderful it is to come when the "Come Here" command is given. For example, take the kitten's favorite toy and drag it with the

A mixed breed cat exhibiting the alert, attentive nature of its species. Photo by Robert Pearcy.

kitten following it to you or make food preparation sounds (open can, shake kibble, etc.) so your kitten associates the "Come Here" with positive and forthcoming pleasant events. Soon your kitten will learn to come running to its name and the command to "Come Here" as long as it pays off most of the time. Be sure to continue rewarding him for coming, or he may become unreliable in his response.

No more yelling after your cat, since a simple request will bring your kitten running to play or be fed. With a bit of patience, you can also quickly teach your kitten to sit with its hind end and front feet planted solidly on the ground and await the food reward you may give. This also teaches the kitten to pay attention to people, to remain in one place, and not use its front claws to grab what it wants. Kittens must first learn to take a small (able to be immediately gobbled up) food treat gently from your fingers, without the help of their paws, prior to teaching "Sit". If the treat is too large, your pet will place it on the ground to chew on it. Start with a hungry kitten on the floor. Take the tasty tidbit of food and quickly bring it over the cat's nose so he must sit to follow the treat. Do not hold the treat too high or move it too fast or the pet will reach to grasp it with his paws. Say the kitten's name followed by repeating the word "sit" during these maneuvers. When the kitten is quietly sitting with all paws on the floor, quickly bring the little treat to the kitten's mouth with a gentle praise of "Good Sit". Ask for a "sit" prior to all meals.

A little time and patience teaching your kitten its name and to "Come" and "Sit", can pay off during the rest of the cat's life. If your kitten will be bathed in the future, introduce him to gently being washed and offer plenty of food and praise during his early baths. Teaching your kitten to be comfortable in the car can also be accomplished early. *Note*: By not confining your pet to a crate, you take the risk that the kitten will fly through the car and be seriously injured in the event of an accident! In addition, kittens can acccidently scratch you or even crawl under the brake and accelerator peddles while you're driving, precipitating an accident.

NOCTURNAL ACTIVITY AND OVERLY EXUBERANT PLAY

Almost all kittens at one time or another exhibit excessive nighttime (nocturnal) activity and overly-exuberant play, which I call the "kitten crazies"! Many owners find these habits exceedingly annoying, especially when they interfere with one's sleep. Although domestic cats are not truly nocturnal, as they do sleep through many of the

Couch potato cats. Photo by Dr. Patrick Melese.

This kitten is being taught to take a hand-fed treat. Photo by Dr. Patrick Melese.

Incorrect training for the "Sit" command. Don't hold the food treat so high above the cat's head that the animal has to reach with its paws. Photo by Dr. Patrick Melese.

nighttime hours, they do have periods of high activity in the late evening and early morning hours (known as a crepuscular activity pattern). This can be quite disconcerting to the owner who retires early or rises later than his kitten. The pet's activities can take many forms: the kitten may merely be a prowler and quietly walk around, checking out his domain. On the other hand, the kitten may choose this as his favorite time to race around the house at full speed practicing acrobatics. This latter behavior will cause most owners and their cats to disagree about what is acceptable nighttime activity. For the kitten who is overly active at night, there are several options. First, getting another well-socialized kitten often helps. Frequently two kittens will play with one another throughout the day and not have as much energy at bedtime. Even if they do get up in the middle of the night, they are likely to play with each other instead of waking up their owners.

Another possibility is to recognize that your kitten has these tendencies and provide it with an energetic play period with you before you retire for the night. If he still insists on trying to wake you up, close your bedroom door. Be careful not to give in to your meowing kitten and let him in, lest you train him to vocalize to get you to open the

Play biting and scratching. Such behavior should be discouraged by the use of appropriate discipline. Photo by Carol Harris.

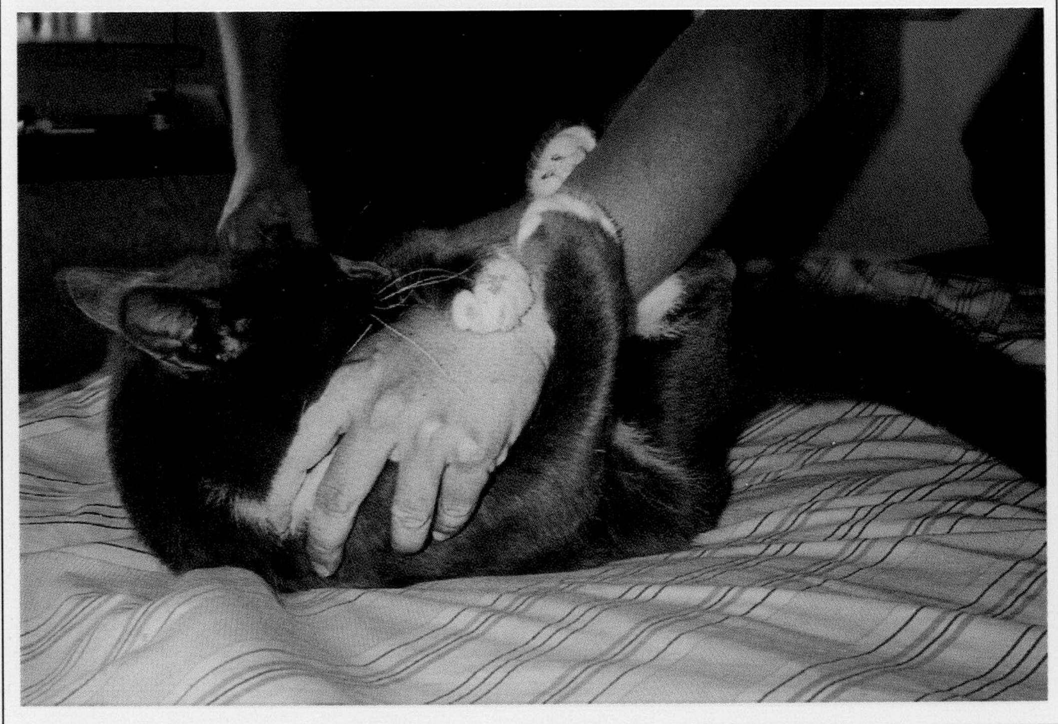

Luring a kitten with a toy to teach the "Come" command. Photo by Dr. Patrick Melese.

door. Try never to reinforce a behavior that you do not want. Frequently, if a behavior no longer gets a response, the kitten stops engaging in it. Nevertheless, you always need to keep your kitten's needs in mind and try to accommodate them as best as you can. In this case, that means realizing what your kitten's play and sleep cycles are. This does not mean, however, that you should totally change your life for your kitten. Your nocturnally active kitten may also be prone to all kinds of overly-exuberant play. This can take many forms, most of which owners find difficult to live with. This kind of play can range from frantic racing around the house, jumping on the furniture and curtains, to simply being overly-exuberant with you and biting and scratching vigorously.

In the case of the normal, highly active kitten, the first thing to do is to put any valuable breakable items away until your kitten learns alternative games (kitten-proof your home). Again, you may wish to consider getting another kitten, since if they have one of their own kind to play and grow with, they turn their enthusiasm toward each other rather than inventing ways to turn your house into a fun zone. The second kitten, not your house, gets the benefit of all that excess energy. If a second kitten is not an option, you must teach your kitten appropriate ways to release his energy. Play hide and

Siamese kittens deciding what to do next. Kittens need an outlet for their energy. Photo by Robert Pearcy.

seek games with him/her and make sure there are plenty of interactive toys and that you use them to play with your kitten. Toys such as the Feline Flyer (sort of a fishing rod with bird feathers on the end of a string), battery-operated balls (especially when placed in a paper bag), fur covered "mice," and many other commercial toys will entertain most felines. Kittens also greatly enjoy chasing after rolled up paper, foil, ping pong balls, playing in paper grocery bags and boxes and many other inexpensive household items. Take a trip to your local pet store and you are sure to find many toys to interest your young cat. Many toys require your participation, and this helps deepen the loving bond between you and your little furry friend.

PLAY NIPPING AND SCRATCHING

If your excitable kitten's play extends to biting and scratching, it needs to be taught the limits of such play. You need to realize that its mother would never allow such liberties, and you do not have to either.

To begin with, try not to play with your kitten directly with your hands unless you are establishing limits in a training session. Instead, use more of the toys which keep people's hands at a distance, such as the ones mentioned earlier in the section above on excess activity. If your kitten still insists on biting and scratching in play, it is time to use a little kitty discipline. Remember, you never need to hit a cat or kitten to discipline it. A mother cat will hiss at her kittens as a form of discipline.

This can also work for you. As soon as your kitten bites or scratches, say "No" firmly and hiss at him loudly. Then, withdraw your hand from his reach. The kitten will instinctively understand the hiss, and by removing your hand, his game no longer works. You may also want to put a bitter-tasting substance (several are available commercially at veterinary offices and pet shops) directly on your hands. This will also help to remind kitty that chewing on you doesn't pay off. In some cases, a quick light swipe on the nose after the hiss can be effective, but stop this immediately if your kitten interprets the maneuver as an invitation for further aggressive play or beomes too frightened.

Other forms of acceptable kitty discipline include the squirt gun or spray bottle, a shaker can, and a small can of compressed air. The shaker can is a soup or soda can into which you've placed eight or ten pennies. You then tape over the opening and you have a very effective, inexpensive noise maker. Compressed air cans are available from camera and computer stores and are very effective due to their air stream and hiss-of-air discharge. It is always a good idea to have several of these devices stationed around the house. They produce an aversion and work only if they are handy when needed and used consistently. When your kitten gets too rough in his play, say a firm "no," hiss and then squirt it with water. These

Kittens at play! Photo by Dr. Patrick Melese.

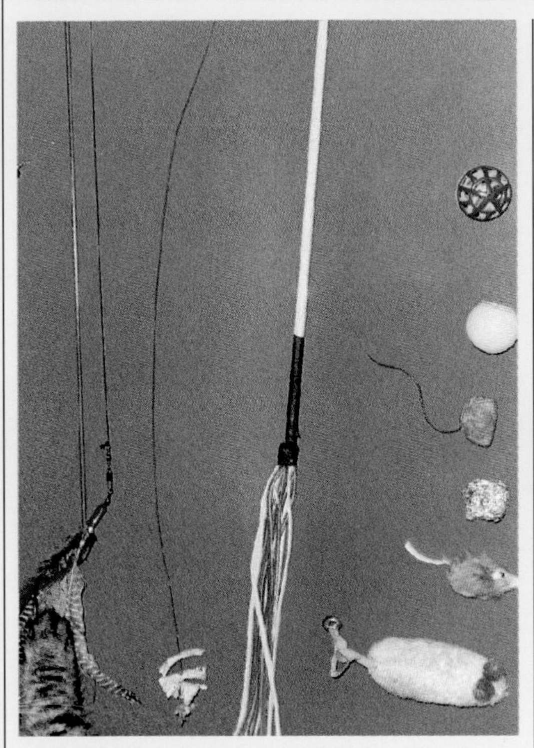

Check all of your cat's toys regularly for signs of wear and tear and replace any that are damaged. Photo by Dr. Patrick Melese.

actions give him the vocal cue that he has overstepped his bounds as well as a physical reminder that there are consequences to his actions. The technique used with the shaker can and compressed air is similar to that of the squirt bottle. Once again your pet is learning that there are consequences to his actions. Keep in mind that although we are discussing problems and forms of discipline, no form of discipline is effective without a heavy counterbalance of love, praise and being taught what is acceptable in the home. You should have a special food reward and praise just as available as your discipline aids. Remember to use these tidbits and praise lavishly when your kitten does something of which you approve. All too often, people remember to reprimand inappropriate behavior and totally ignore or do not reward behavior they wish to see repeated. Kittens, like people, thrive on a combination of reward and appropriate correction. Reward behavior you wish to continue, and your chances of success will increase dramatically.

This looks like fun! Photo by Dr. Patrick Melese.

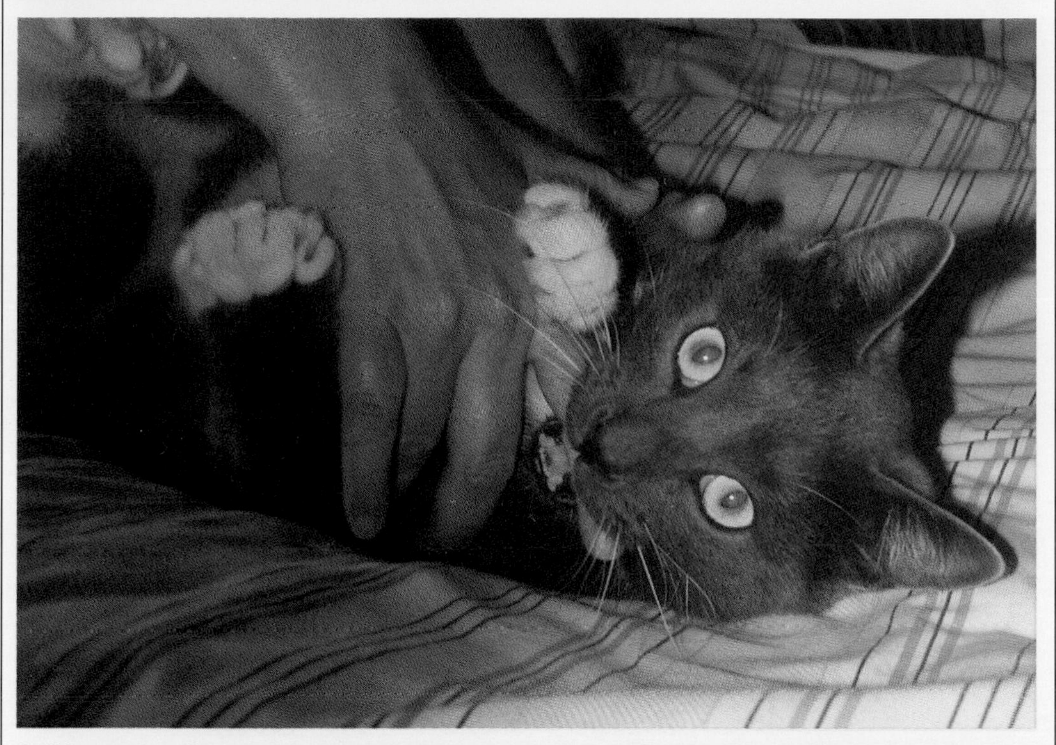

Cat bites can hurt. Some people make the mistake of encouraging their kittens' play biting behavior because they think it's cute, but when their pets grow up, they feel quite differently. Photo by Carol Harris.

HANDLING/PETTING AGGRESSION

Some kittens can become rather aggressive after seemingly enjoying being petted for a period of time. This type of kitten may be lying quietly in your lap soaking up the attention and enjoying your touch, even purring, when it suddenly turns to scratch and bite you. It will usually immediately jump off your lap and quickly move away from you. This occurs most frequently in male cats. There are several theories as to why domestic cats do this; one is that while the cat does enjoy the petting, it reaches a threshold whereby he no longer finds being touched acceptable. Regardless of why the kitten exhibits this behavior, the bottom line is that it is unacceptable to be bitten or scratched when you are sharing a pleasant quiet moment with your cat. Punishment in these circumstances is usually ineffective and can act to damage the bond between you and your cat. It is far better to try and modify the behavior to be more suitable for both of you.

When stroking your pet, watch closely for any signs of agitation. A twitching tail, flattening of the ears, or tensing of the body could be signs that your cat is reaching the level of intolerance to petting. If you see signs, however subtle, that your kitten is no longer enjoying being petted, stop, move your hand and face away from his reach and let him walk off. You can even get up so that he must jump off of your lap. It is far better to prevent the behavior than to try and punish it.

When your kitten is quietly enjoying being petted, try rewarding him with occasional small food treats. Try waiting longer, and longer in the petting session prior to rewarding the kitty. This can help extend the amount of time he permits petting and helps him enjoy the prolonged stroking.

Remember not to push your kitten beyond what he's able to tolerate and to reward his quiet acceptance of handling. Remind all family members to handle him gently and to respect his needs. With patience, you will all be able to live with this problematic behavior in your young cat.

CLIMBING ONTO COUNTERS AND FURNITURE

To the chagrin of most owners, they soon find that their kittens cannot resist exploring every available surface. You must keep in mind that the way a kitten explores his environment is to physically encounter each and every square inch. Cats need to step on, scratch, rub against, sniff, taste or otherwise contact everything they come across. Furthermore, remember that cats are climbers by nature.

Even adult cats will climb onto counters amd furniture if they are not properly trained. You may want to cat proof any areas that contain valuable fragile objects. Photo by Dr. Patrick Melese.

Cats are natural-born climbers that enjoy surveying their surroundings from above. Photo by Robert Pearcy.

This cat has earned a reward for a "good sit." Photo by Dr. Patrick Melese.

Keep training sessions reasonably short. If your cat seems bored or disinterested, wait until later in the day to work with him. Photo by Dr. Patrick Melese.

If you have a kitten, you are likely to find him in some surprising and not necessarily desirable places. You should decide early where your kitten will be allowed to go and where will be off limits. It is absolutely necessary that the entire family be aware of these decisions and agree to enforce them consistently. If your kitten is not going to be allowed on the kitchen counters or table, then no one can permit it to be up there. The best way to sabotage your plan is to have a family member inadvertently reward the kitten for jumping on these excluded areas by offering a bit of food, or petting him because he is so cute. The problem can also be compounded when food is left on the counter, inviting bad behavior. Instead, discourage kittens from getting onto areas considered "off limits" by making the area itself aversive. The less direct punishment (punishment that seems to come from you) the more likely your cat is to stay off counters or furniture at all times, not just when you're in the room. For example, if every time that your family sees the kitten in an inappropriate area they consistently reprimand him with an effective "NO, HISSSS" and even follow up with the aversions mentioned above (water, can, air), the kitten will likely learn to stay off of these surfaces when someone is in the room. The young feline often learns that he can comfortably use these areas when no one else is around. This is likely since cats in nature are accustomed to using parts of their environment when others are not around. (Time partitioning: Instead of territorial space being divided up, the times of day these areas are used by a certain animal are divided up, which allows several animals to use the same space at different times and not encounter each other directly.) Using your imagination, you can come up with a number of harmless ways to discourage your kitten from getting onto prohibited surfaces. On counters, or any smooth surface, place strips of tape with adhesive on both sides (double-stick tape, or lint roll tape). A kitty that jumps on these surfaces will find the stickiness on his paws very unpleasant to walk on. After a few times on these sticky surfaces most kittens will decide it is not worth getting up there again. This change of behavior gives you the opportunity to praise and reward your kitten when you see him in the room and not up on the excluded surface. Do not omit this praise and reward phase for good behavior. Remember that kittens quickly learn from positive reinforcement.

Another method of teaching kittens is to use devices that create noise when the kitten gets up on the forbidden item. Among these devices are Snappy Trainers (a sort of

Turkish Angora photographed by Robert Pearcy.

safety mouse trap), shaker cans, commercial electronic motion detector noise devices (e.g., Tattle Tale, Scraminal, etc.), or commercial electrostatic shock pads (e.g., Scat Mat). Other than shaker cans, Snappy Trainers are among the most cost effective. Snappy Trainers are a commercially available device that resembles a mouse trap with an attached large plastic paddle. This plastic piece prevents injury to your kitten and allows the Snappy Trainer when tripped to fly up into the air and startle your kitten.

Shaker cans can be set up in such a fashion as to ensure that your kitten will knock them off when it jumps on the counter or table. They can create an incredible clattering noise, which effectively startles many kittens. Stacked, empty cans booby trapped to fall if a kitten walks by can create a similar effect. Commercial motion-sensing devices are small battery-operated units that emit high-pitched noise when anything crosses their motion sensing beams (Scraminal), or when vibrations are detected (Tattle Tale). The Scat Mat produces a harmless but uncomfortable electric shock when the kitten steps onto its surface. These items are somewhat more expensive than the other methods we have suggested; however, they are highly effective on many kittens.

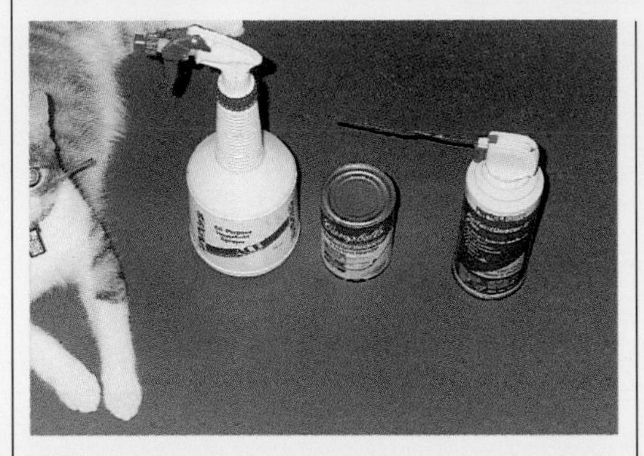

The items shown here can safely be used as deterrents when your pet is exhibiting undesirable behavior: plant sprayer, shaker can, and compressed air cannister. Photo by Dr. Patrick Melese.

Most of these techniques are equally effective on furniture. Be cautious with double stick tape, as it can be hard on some upholstery. You can, however, pin aluminum foil onto your furniture for a similar aversive effect. In addition, many products are marketed to discourage your kitten from going where you don't want him to be. They come in liquid, gel or spray form. They have an unpleasant taste or odor to deter the kitten. They can be effective only if your kitten finds them aversive. Some kittens are not put off and may actually like them. Often a plant sprayer or squirt gun can be effective in these cases, but the kitten must not know that the water comes from you. If he realizes where it comes from, he will learn to stay off when you are close by, rather than at all times—thereby creating an owner-absent problem (problem behaviors that only occur when people are not around. Usually occurs when the

Nothing escapes the scrutiny of a cat. Photo by Robert Pearcy.

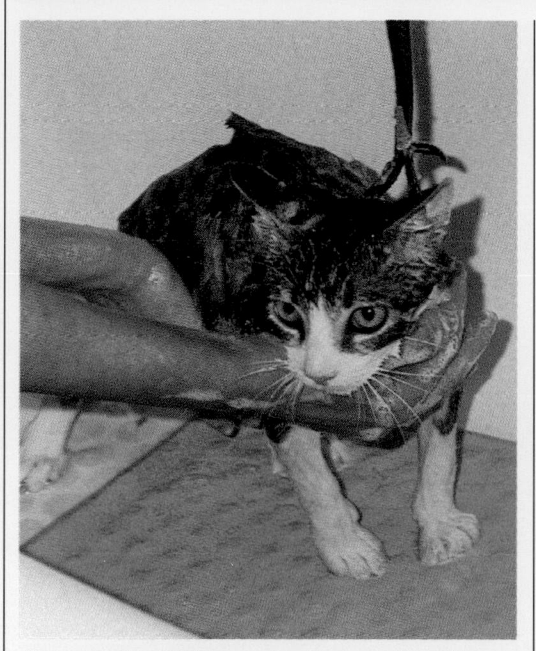

Above: First bath! A new experience such as this can be upsetting, but this kitty seems to be doing just fine. Photo by Dr. Patrick Melese.

pet associates punishment for undesirable behavior with people.)

Whichever method or combination of methods you choose, always keep in mind that you do not want to harm your pet, but merely to convince him that getting up on counters or furniture is a bad idea.

Most importantly, never forget to praise and reward your kitten when he has resisted the temptation to go where he shouldn't.

SATISFYING YOUR CAT'S NEED TO CLIMB

The final subject that we will address in this chapter is climbing on furniture. As I

Below: There are certain areas of a home that should always be off-limits to a cat. Photo by Dr. Patrick Melese.

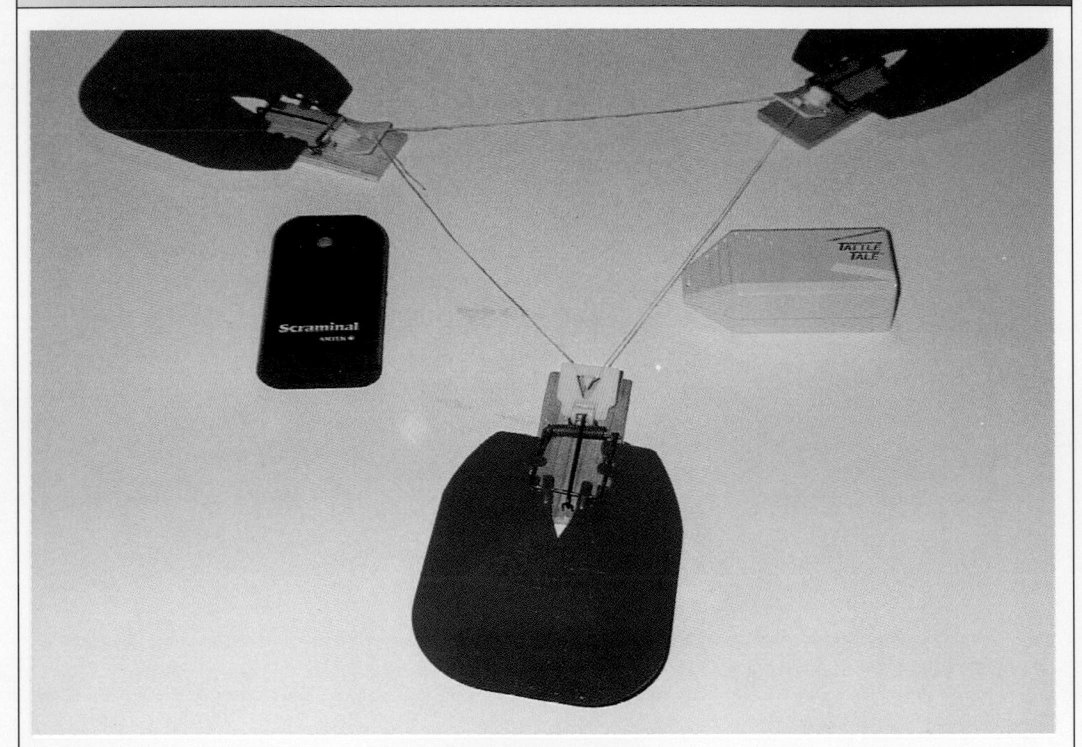

Above: You can purchase various types of "booby-traps" to discourage your cat from going where it shouldn't go. Photo by Dr. Patrick Melese.

Below: When they are stepped on, Snappy Trainers™ fly up into the air, thus startling a kitten and prompting it to leave the area. Photo by Dr. Patrick Melese.

Above left: Scat Mat™ delivers a mild electrical charge that deters the approach of most cats. Photo courtesy of Contech. *Above right:* This cat is about to climb his cat tree. The rope that encircles the trunk affords plenty of opportunity for the cat to work his claws. Photo by Dr. Patrick Melese. *Below:* There are products specially designed to keep cats and other household pets off of furniture. Photo courtesy of Dr. Gary Landsberg.

Three little kittens getting into mischief. Photo by Robert Pearcy.

mentioned earlier, cats love to climb. They enjoy being in high places. This is another requirement that you should satisfy. Providing your kitten with his very own tree can not only make kitty happy, but it can also save "off limit" items in your house from being climbed upon. A dedicated climbing area also gives your kitten a place to exercise, which will keep him healthy and prevent him from becoming overweight. There are many commercially available cat trees on the market. They vary in design as dramatically as the homes for which they are purchased. Some considerations to keep in mind are: (1) will the kitten like it? (2) is it durable and well made? (3) is it stable? If the kitten doesn't like it, it will not be used and you will just waste money for a piece of furniture you have no use for. Many kittens enjoy climbing trees that span floor to ceiling and are partially made of real tree branches. The ones that have resting platforms at several levels are attractive to most cats. Varied textures make these climbing trees even more attractive to the kitten. Often you can find them covered in combinations of carpet, sisal rope and bare wood. Teach your kitten to use this post as a scratching post as well as a climbing and resting platform, and you will help save both your furniture and great deal of aggravation.

Train your pet with lots of love and patience and a little discipline and your kitten will grow into the friendly, well-behaved companion with whom you want to live for many years.

SUMMARY

Kitten behavior problems are one of the most common reasons for owners to seek a behavioral consultation. There is no doubt that a kitten that misbehaves puts a lot of stress on the owner-pet bond. This chapter will help you train your kitten right the first time, and also offers some tips for kittens who need some extra assistance.

ADDITIONAL READING:

Bohnenkamp, G.: *From The Cat's Point of View* (book) and *Kitty Kassettes* (audio cassettes). James and Kenneth Publishers, 2140 Shattuck Ave. #2406; Berkeley, CA 94704 (510) 658-8588.

Fox, M.W.: *Supercat*, Howell Book House, New York, 1990.

Loxton, H.: *The Noble Cat.* Portland House, 1990.

Milani, M.M.: *The Body Language and Emotion of Cats*, William Morrow & Company, Inc., 1987.

Turner, D.C.; Bateson, P. (Eds): *The Domestic Cat: The Biology of Its Behaviour*, Cambridge University Press, 1988.

Dr. Hunthausen is a pet behavior therapist who works with people, their pets and veterinarians throughout North America to help solve pet behavior problems. Since graduating from University of Missouri School of Veterinary Medicine in 1979, he has practiced in the Kansas City area. He is director of Animal Behavior Consultations, which provides behavior consultations as well as training services. He writes for a variety of veterinary and pet publications and is coauthor of the textbook, The Practitioner's Guide to Pet Behavior. *Dr. Hunthausen frequently lectures on pet behavior and currently serves on the behavior advisory board for the following journals: Veterinary Forum, Feline Practice, Canine Practice and Veterinary Technician. He is a member of the American Veterinary Society of Animal Behavior and has served on its executive board as secretary-treasurer and currently as its president. Besides hosting his Kansas City radio talk show, "Animal House," he is a regular, featured guest on the "The Morning News" for the local NBC television affiliate.*

Feline Housesoiling Problems

By Wayne Hunthausen, DVM
Animal Behavior Consultations
4820 Rainbow Blvd.
Westwood, KS 66205

INTRODUCTION

Why does a normally well-behaved cat suddenly start eliminating outside its litterbox? Is the pet trying to send a message or retaliate for some perceived injustice? Is it jealous because the owner is spending more time with a new baby or friend? Actually, none of the above are reasonable causes of a failure to hit the box. While there are literally thousands of reasons why the pet might decide to eliminate in areas away from the litterbox, most housesoiling problems fall into one of two categories: inappropriate elimination or spraying. Inappropriate elimination is the deposition of urine and/or stool in an unacceptable area of the house on a horizontal surface. Spraying occurs when the cat backs up to an upright object and squirts urine on it. During this latter behavior, the cat will usually stand tall and hold its twitching tail in an upright position. This is a marking behavior.

INAPPROPRIATE ELIMINATION

Since medical problems are common causes of inappropriate elimination, it is very important that the pet is examined for signs of disease that might be an underlying cause of the behavior problem. Any discomfort that is associated with the litterbox may cause the pet to avoid it. A cat that has inflammation of the lower urinary tract or colon will usually experience pain during the act of elimination. The pet may think that the box has something to do with causing the pain it experiences. In an attempt to avoid the pain, it will look for new places to eliminate. As long as the pain exists, the cat will continue to pick new places in which to urinate or defecate.

Problems that result in excessive urination, such as diabetes or kidney failure, can also lead to housesoiling problems. Some of these cats produce so much urine that the litterbox is constantly soaked and frequently too damp to be acceptable.

Other causes of inappropriate elimination fall into the two broad categories of environmental factors that cause the cat to avoid the box and factors that attract the cat to areas away from the box.

FACTORS CAUSING LITTERBOX AVOIDANCE

Litter Factors

The feel of the surface texture against the paws seems to play an important roll in influencing where a cat chooses to eliminate. Most cats are born with an innate desire to eliminate in substrates that can be scratched and scooped in order to form a shallow depression for elimination. While some individuals show rather general preferences, others can be quite picky. If the particle size, texture or depth of litter is not exactly right, the pet will avoid the box. The odor of the litter can sometimes cause problems. Some cats seem to dislike the fragrance added to certain commercial litters to mask odors.

If the pet suddenly starts urinating and defecating away from the litterbox when the type of litter is changed, it is highly likely that the new litter is unacceptable. If you believe the pet is housesoiling because it doesn't like the litter you are offering, you'll need to switch back to a previously used litter or try to find something new that it likes. In attempting to locate an acceptable litter, it is usually safer to offer different types in a second or third box rather than continually changing brands in a single box. Clumping litters are usually a good choice since some studies suggest that they may be preferred by most of the general cat population. For the pet that won't use any type of commercial litter, you'll need to explore other substrates. Try sand, potting soil, crushed leaves, straw, peat moss, wood shavings or

Closing window shades prevents the indoor cat from watching outdoor cats, which can trigger it to spray indoors. Photo by Dennis Bastian.

combinations of these substances until you find some substrate in which the pet will eliminate. Then, very gradually replace that substrate with plain clay litter.

Some consideration should be given to how much litter is placed in the box. Some cats are picky about the depth of the litter. If it's too deep or too shallow, they may avoid using the box. A two-inch depth is a good starting point when filling the box.

In extreme cases where the pet refuses to eliminate in any type of substrate in the litterbox, it may be necessary to confine the pet to a large cage or small room covered with litter. An upside-down box or shelf can be provided as a resting place above the litter. The pet should frequently be allowed out of the confinement area, but only after it has eliminated in the litter, and should be watched very closely when it is out of confinement. This may be required for many weeks or months before the amount of litter on the floor can gradually be decreased to cover the small area of a litterbox.

Litterbox Factors

The litterbox must be cleaned often enough to prevent the pet from avoiding it. How frequently you must clean it depends on the individual pet. Some cats require daily cleaning to keep them happy, while others will tiptoe around until every last square inch of the box is used. If you're not sure how often your cat likes to have its box cleaned, begin by scooping it at least once daily and completely cleaning the box at least once weekly. If necessary, you can gradually try decreasing the frequency of scooping. If you are too busy to keep the box cleaned as frequently as necessary, then try providing an additional box.

Sometimes, cleaning the box can actually cause the cat to avoid it. This occurs when harsh-smelling disinfectants are used. The cat's nose is many times more sensitive than ours, and it may be put off by odors that we can't even detect. For the pet that is especially sensitive, problems can be avoided by using scalding hot water to clean the box. Another solution is to use two boxes and allow one box to air dry after cleaning while the other is put in use.

The size of the box can be an important consideration for some cats. If it is too small, the pet may either avoid it or hang its rear-end over the side of the box when it eliminates. Providing a bigger box or one with higher sides will usually take care of this problem.

Suddenly switching to a covered litterbox can occasionally cause problems. If you want to use a covered box and you're concerned that the cat might not accept it, try covering a regular litter box with a large cardboard box with an open end and gradually change to smaller boxes until the size of a commercial covered box is reached.

If you started using a plastic liner just prior to the appearance of the problem, that could be the cause of the housesoiling. Some cats are put off when they snag their claws in liners.

At all costs, you want to avoid associating anything the cat doesn't like with using the box. Although it might be convenient to medicate the pet when you catch it in the box, it will quickly learn to stay away from the box in order to avoid being caught there. Children should be taught not to disturb the pet in the box. If the kids won't listen or if the pet dog bothers the cat when it is using the box, placing the box up high–where it can't be reached–will help. Another solution is to place the box in a room that has a cat door inserted into the door to the room or a baby gate in the doorway so that the cat can enter, but the kids and/or the pet dog can't.

Litterbox Location Factors

Too much activity in the area where the litterbox is kept may cause the pet to avoid it and seek quieter areas. Some cats need more privacy than others and should be provided with a litterbox in a quiet, relatively secluded area. Another reason that a cat may avoid an area where the litterbox is kept is if some frightening event has occurred in that area. If the box is kept in the laundry room and the pet is using it when the washing machine goes off balance, it may not return to that room. I once treated a cat for housesoiling that began avoiding the room where the box was kept after some books shifted and fell off of the bookshelf in the room while the pet was in the litterbox.

FACTORS THAT ATTRACT THE CAT AWAY FROM THE LITTERBOX

Surface Preference

The cat's preference for eliminating on specific surfaces is determined by genetics and learning. As mentioned, most cats are born with an innate desire to eliminate on surfaces that can be scratched to provide a shallow depression for elimination. Variations do occur, and an occasional cat will have more general preferences for elimination and will eliminate on a wide variety of surfaces.

A common surface-preference problem occurs when the pet eliminates in the natural soil surface of a potted plant. An easy way to correct this problem is to change the surface by placing decorative moss, bark, gravel or other types of surface cover over the soil.

On rare occasions, a pet may choose unusual surfaces, such as paper, or hard surfaces, such as tile or linoleum. If your cat prefers hard surfaces for elimination, try confining it for a period of time in a small room that is covered with old sheets, towels or other types of fabric where the only hard surface available is a large, empty litterbox. Very gradually add a fine type of litter to the box until the pet becomes accustomed to eliminating in a litter-filled box.

It has been my experience that

The litterbox should be placed in an area where the cat will not be disturbed by other household pets. Photo by Dennis Bastian.

some cats that start urinating in sinks, bathtubs or similar areas have lower urinary tract disease. Exactly why they do this is unknown. If urinary disease is the problem, you may also note that small amounts of urine are frequently voided, the pet strains when it voids and the urine has a pinkish, blood-tinged color.

It is not uncommon for new surface preferences to develop when the cat avoids the litterbox because of aspects about it that are unacceptable. For example, if the owner changes to a brand with an odor the pet does not like, it may start eliminating on the living room carpet in an effort to avoid the unacceptable litter. In the course of time, it may decide that the carpeting is just as good a place to scratch and eliminate as litter, thus developing a new surface preference for eliminating. Cats may also develop preferences for eliminating on plastic, paper, wood or clothing. Some of the cats that pick a specific family member's clothing or favorite chair on which to repeatedly eliminate may do so because of some type of anxiety that relates to that person, such as inappropriate punishment or absences from the home.

Location Preference

Just as new surface preferences tend to develop when the pet becomes unhappy about the litterbox, so can new location preferences develop. The pet may find that is prefers to eliminate in a more secluded or quiet area of

Some cat owners put a cat door in the door to the room where the litterbox is kept. This ensures that the cat can use its litterbox without being bothered. Photo by Dennis Bastian.

the home. The habit of eliminating in a certain area while the cat is avoiding the litterbox or litterbox area can become quite strong. This strong location preference will keep the pet coming back to the same area to eliminate even after all urine odor has been removed from the area, medical problems have been treated and a favorable litter or litterbox has been provided. Placing a litterbox at the site of housesoiling and then, gradually moving it to an acceptable area may work, but usually takes time and patience.

MISCELLANEOUS FACTORS CAUSING INAPPROPRIATE ELIMINATION

Anxiety

Anxiety can be an underlying cause of housesoiling. A fearful stray that has been brought into the home may not leave secluded hiding areas to get to the litterbox. If a pet has been harshly punished by a family member, it may hide in out-of-the-way areas to avoid being caught and punished. This may prevent it from getting to the box when it needs to eliminate.

Another cause of anxiety that occasionally triggers inappropriate elimination is separation anxiety. A cat that is very close to a family member may urinate outside of the box as the suitcases come out, while the owner is gone, or for a short period after the person returns home. The easiest solution for this kind of problem is to keep the pet confined to a relatively small room during periods when it is likely to miss the box.

Anti-anxiety medication may temporarily be prescribed by your veterinarian if anxiety is an important factor contributing to your cat's housesoiling problem.

Marking

Although most marking behavior occurs in the form of spraying urine on upright objects, a rare cat will mark territory by urinating in a squatting position. This can be a difficult type of problem to diagnose. Keeping a diary to track common times at which the pet urinates on horizontal objects and correlating these times to environmental changes may help uncover the motivating factors. Treatment of horizontal marking behavior is the same as for spraying behavior.

Geriatric Problems

As the pet becomes older, weakness and arthritis may make it difficult to reach a litterbox that is kept in the basement, on an elevated platform or at the far end of the home. Getting over the high sides on the litterbox may become a formidable task. You should have several litterboxes with low sides available in areas that are easy for the pet to reach.

In some elderly cats, housesoiling can be the result of deteriorating mental function. These pets are often disoriented and just can't seem to remember where they are, much less where the box is located. There is no treatment for this type of problem, and the only solution is to keep the pet confined to an area of the home that can easily be cleaned.

REESTABLISHING LITTERBOX USE

Besides uncovering and correcting the underlying causes of the housesoiling problem, there are two important things that can be done to promote the habit of using the box on a regular, consistent schedule. First, you should reward the pet every time you see it using its box. Praise the behavior and then immediately give a food reward when the pet steps out of the box. The second consideration is to provide close supervision or confinement. The pet should be confined to a small area with a litterbox and allowed out only when it can be supervised 100% of the time. When confined to a relatively small area such as a bathroom or utility room, most cats seem to prefer to eliminate in the box rather than soiling the floor. If the cat still refuses to use the litterbox when confined to a small room, the confinement area should be changed to a large cage. It's then a matter of confining it long enough for a consistent habit to become established. This may take several weeks or more. Once the cat has used the litterbox in a confined area for an extended period of time, you can begin to gradually allow more freedom in the house.

Don't forget about the pet during the confinement period. Be sure to get it out for frequent play and social sessions. The object of confinement is not to punish or socially isolate the pet, but to help prevent it from making any mistakes.

PREVENTING THE PET FROM RETURNING TO THE SCENE OF THE CRIME

Once the cat has more freedom, there is a risk that it may return and eliminate on previously soiled areas. The pet can be taught to avoid these areas by placing safe booby-traps, moth balls, mentholatum, citrus sprays or lemon rinds on the area. You may need to experiment. Each cat is an individual. Something that repels one cat may attract another.

Most cats won't eliminate where they eat, sleep or play. Placing food bowls, water bowls, the pet's bed or toys in the area may inhibit elimination. Objects that can be placed over areas to protect them include plastic sheets, foil, upside-down plastic carpet runner, double-stick carpet tape or a piece of furniture. An inch of water in the bathtub or sink will curb elimination there. In some areas, such as the corners of the basement, it may be prudent to place a litterbox where the cat has been soiling. If the pet is eliminating inappropriately in just one room, the easiest solution may just be to keep the door closed.

Removing the odor is important, although it is usually not enough to remove the odor without doing anything else to the area. Cats will return to an area due to the formation of new surface or location preferences as well as being attracted by odor. Products that are formulated specifically to work on feline stool and urine odors should be used. These products need to make contact with the organic material. An ample amount should be poured on carpeting and porous surfaces, rather than lightly sprayed. A 50:50 mixture of white vinegar and warm water will do a satisfactory job if nothing else is available.

SPRAYING

Spraying is a marking behavior that is likely to occur when the pet feels as if its territory is being invaded, or if there is something in its life that makes it anxious. Neighbor cats visiting in the yard or too many cats in the home can trigger this kind of problem. The incidence of spraying increases from 25% in single-cat households to 100% in households with more than 10 cats. Unneutered males or females in heat are the individuals most likely to spray, although some neutered cats will take up this nasty little habit. Studies have shown that as many as 10% of castrated male cats and 5% of spayed female cats take up spraying on a frequent basis as adults. The objects that are commonly sprayed include doors, walls by doors or windows, new objects in the house, and furniture.

The tendency to spray is influenced by factors pertaining to the individual (hormones, personality), things in the environment that are upsetting to the cat (new roommate, new cat in

Facing Page: Longhaired mixed breed. Photo by Robert Pearcy.

the neighborhood, remodeling, moving) and its relationship with the owners (change in the work schedule, absences from home, spending less time with the pet, inappropriate punishment). Sometimes, just the suggestion that another pet has invaded its territory can cause the pet to spray. For example, if items brought into the home (such as clothing, suitcases, firelogs, plants) have the odor of another animal, they may trigger spraying by the pet.

TREATMENT OF SPRAYING

The two main approaches to correcting spraying behavior include getting rid of the stimuli for the behavior and changing the cat's normal response to those stimuli.

Controlling the Stimuli for Spraying

If outdoor cats are the reason for spraying, the owner should consider discouraging their visits by using a water hose or humane booby-traps. Sometimes sympathetic neighbors can be talked into keeping their cats indoors. Anything in the yard that might attract roaming cats should be removed (bird feeders, garbage, food). Besides removing the stimuli, the owner can remove access to the stimuli. The spraying cat should be kept out of windows or out of rooms that permit it to view outdoor cats. Drapes can be closed. Window sills can be modified so the pet cannot perch on them. Chairs near windows on which the cat sits can be moved. What the cat doesn't know about what is going on outdoors won't hurt him—and may save his life as well as protect your happy home.

Urine odor should be cleaned from the outside around doors and windows. Some individuals will spray less indoors if they have more access to the outdoors. Other do better if kept inside more. If other pets in the household are contributing to the problem, they should be separated. In some cases, reducing the number of pets in the home may be the only solution.

Changing the Cat's Response to Stimuli for Spraying

Neutering is very successful in curbing spraying because it reduces the pet's response to stimuli in the environment that trigger marking behavior. In most cases, neutering should be done as soon as possible. As previously mentioned, ninety percent of males and ninety-five percent of females will not spray following castration or spaying.

Medication is often required to control spraying behavior so you may need to discuss the problem with your veterinarian. In the past, progestin hormones were commonly prescribed for the spraying cat, but they can have serious side effects. Safer medications are now available, and hormones are infrequently used for this type of problem.

Some cats may be tempted to use houseplants as their litterboxes. Photo by Robert Pearcy.

Persian kitten photographed by Robert Pearcy.

Drugs do not work for every cat, and all medication can have some side effects, so be sure to discuss expected behavioral changes and side effects with your cat's veterinarian prior to starting therapy. If the medication will be effective for your pet, you usually can expect a decrease in the marking behavior within a few weeks of starting drug therapy. Once some solid progress has been made, your veterinarian will probably want you to slowly decrease the dosage.

PUNISHMENT

Punishment is the least effective tool for correcting housesoiling. Punishment may actually make things worse if anxiety or fear is an important component of the problem. Under no circumstances should you ever swat or physically punish the pet. Rubbing the pet's nose in the mistake is a definite no-no, as is roughly handling the pet and placing it in the litterbox. If you catch the pet in the act of eliminating in an inappropriate area, squirt it with a water gun, toss an object near it, or make a loud noise so it will learn to associate something negative with the behavior. Any type of punishment should be given only during the behavior or within one to two seconds after the behavior ceases. If you don't catch your pet in the act, don't even scold it. Remember that it's very important that your cat not associate the punishment with the person doing the punishing.

SUMMARY

Be sure to start working on the problem right away. Poor litterbox habits rarely get better on their own and short-term housesoiling problems are always easier to correct than long-term ones. Remember to have your veterinarian help rule out underlying medical problems.

Housesoiling problems can be complex, difficult problems to correct, but most of the cats involved can be guided back to their normal, good habits with a little detective work to uncover the contributing factors and lots of patience.

ADDITIONAL READING

Bohnenkamp G. *From the Cat's Point of View.* San Francisco: Perfect Paws, P.Box 5214, San Francisco, CA, 94188, 1991.

Bradshaw J.W.S. *Behaviour of the Domestic Cat.* Melksham, UK: Redwood Press Ltd, 1992 .

Campbell W. *Better Behavior in Dogs & Cats.* Goleta: American Veterinary Publications, 1986.

Neville P. *Do Cats Need Shrinks? Cat Behavior Explained.* Chicago: Contemporary Books, 1990.

O'Farrell V., Neville P. *Manual of Feline Behaviour.* Cheltenham: British Small Animal Veterinary Association Publication, 1994.

Turner D.C., Bateson P. (ed.) *The Domestic Cat: The Biology of Its Behavior.* NY: Cambridge University Press, 1988.

Dr. Lowell Ackerman is a nutritional consultant in addition to being board-certified in the field of veterinary dermatology. To date, he is the author of 13 books and over 150 articles dealing with pet health care and has lectured extensively on these subjects on an international schedule. Dr. Ackerman is a member of the American Academy of Veterinary Nutrition and the American Veterinary Society of Animal Behavior.

Nutrition and Behavior

By Lowell Ackerman, DVM, PhD
Mesa Veterinary Hospital, Ltd.
858 N. Country Club Drive
Mesa, AZ 85201

INTRODUCTION

There are many ways in which diet and behavior can be linked, and it is only recently that these have begun to be properly explored. It now appears that some problems may have a nutritional basis, while others are responsive to nutritional manipulation. It is not outlandish to assume that some cats might have behavioral problems related to their diet. After all, many cats eat high-calorie, high-protein commercial diets, liberally laced with additives, flavorings, preservatives and other processing enhancements. All of these features have become suspect by different investigators pursuing different aspects of behavior problems. This discipline is still in its infancy and not immune to controversy. However, let's take a look at some of the ways that nutrition and behavior may be linked.

EFFECT OF DIETARY INGREDIENTS ON BEHAVIOR

Most veterinary behavior specialists have considered the role of protein in behavior problems. Both the quantity of protein and its quality and extent of processing have recently become suspect. High-meat diets can result in lowered levels of the neurotransmitter serotonin in the brain, which can make some animals more aggressive. Conversely, high-carbohydrate diets result in higher levels of serotonin in the brain. Can feeding high-carbohydrate diets be an effective treatment for some forms of aggression? This concept is being explored, but clear-cut answers are not yet available. The topic of feline aggression is so complicated that it is difficult to make any generalizations.

Casomorphine is derived from the digestion of casein and exorphines from the digestion of gluten. Together with hormones, hormone-like substances, and pheromones naturally present in many cat foods, all have been shown, scientifically, to alter normal animal behavior. Casomorphine and the exorphines, which would be provided by milk proteins and cereals respectively, can trigger behaviors in cats not unlike giving them morphine or similar substances. The overall effect of casomorphine and exorphines

Could behavior be linked to diet? Photo by Dr. Gary Landsberg.

from commercial cat foods has not been adequately investigated.

Another concern is that some problems are more common in cats fed dry food than those fed a canned ration. Some people blame moisture content, but there is probably a better explanation. Most dry cat foods are heavily preserved with antioxidants so they can last on store shelves for months without going rancid. Since canned diets are heat sterilized before packaging, additional use of preservatives is not needed. Some of the preservatives getting increased scrutiny are ethoxyquin, butylated hydroxyanisole (BHA) and butylated hydroxytoluene (BHT). Only recently has any scientific research been directed in this area.

Testing the Theory

It's not enough to make claims and hypotheses about health care issues. There must be a way of proving or refuting such theories for individual animals. This is possible to a limited extent, as long as certain provisions are accepted. The hypothesis of high-protein or preservative-rich diets contributing to behavior problems can be easily tested by feeding a high-quality but low-protein diet for as little as seven to ten days. This time frame is not long enough to uncover adverse food reactions (up to 12 weeks would be necessary) but is sufficient for judging the impact of protein and preservatives on pet behavior. The diets must be homemade protein sources that are suitable, including boiled chicken, lamb, or rabbit combined with

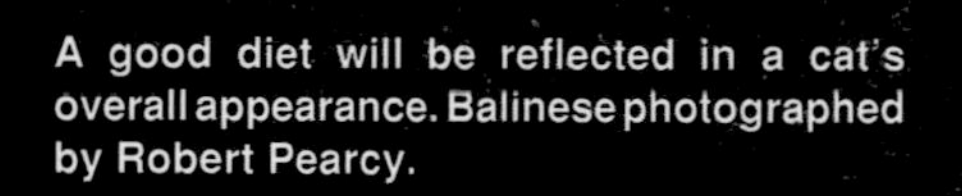

A good diet will be reflected in a cat's overall appearance. Balinese photographed by Robert Pearcy.

boiled white rice or mashed potatoes. This also limits problems that might occur from cereal-based diets (e.g., exorphines), milk proteins (e.g., casomorphine), and preservatives. The meal should be mixed as one part meat to three parts carbohydrate and fed in the same amount as the pet's regular diet. Only fresh water should be provided during the trial. No supplements, treats or snacks should be given. This diet is not nutritionally balanced, but that should make little difference for the seven to ten days in which the trial is being conducted. However, it is critical that cats not be maintained on these diets, certainly not for longer than ten days.

If there is a response to the diet trial, it is advisable to follow up on the process by challenging the pet with specific potential offenders. The first test would be to increase the protein component of the diet (50:50) to see if there is a behavioral change. If so, this helps confirm the suspicion that it is the protein component that is contributory. Adding commercial foods or treats with specific preservatives is the best way of determining the role of food additives in the problem.

What to Do About It

For animals that respond to a homemade low-protein, preservative-free diet, there are many options available. Regular use of a homemade diet should be discouraged unless a completely balanced ration can be formulated. Low-protein diets are commercially available and are the most convenient option. Remember that low protein should not mean low quality. Look for diets with high-quality protein in moderate amounts and an easily digested carbohydrate source. Cereal-based diets tend to have a lot of exorphines, which may contribute to behavioral problems. Start with canned diets, which tend to have few if any preservatives. Dry foods have the most preservatives, and semi-moist foods have too high a sugar content. If the conditions worsen when the pet is put on a commercial ration, there are likely more problems than just protein content to consider. Home-delivered, preservative-free, home-prepared and frozen pet foods are all options with which the veterinarian should be familiar.

For cats with reactions to preservatives, canned foods are an option; and there are also preservative-free diets commercially available. Both of these are usually acceptable, but current regulations make it almost impossible to be assured that there are actually no preservatives in preservative-free diets. Manufacturers need only list on the label those preservatives that they add during ration preparation. However, there is no guarantee that the manufacturer did not purchase the raw ingredients already preserved. If a pet responds well to the homemade diet, and challenge feeding fails to uncover a culprit, consider additives as a likely candidate. When commercial diets cannot be used, homemade diets remain a final option. At this point, it is worth having a diet recipe prepared by a nutritionist to ensure that nutritional requirements will be met. Alternatively, computer software is available (e.g., Small Animal Nutritionist: N-squared Computing) so that custom diets can be formulated by veterinary practitioners.

FOOD ALLERGY/INTOLERANCE

It is not unusual for pets to react adversely when fed certain foods. Occasionally, diet-related problems can affect systems other than the skin and digestive tract, and behavioral problems have been reported. Some animals can even develop diet-related seizure disorders.

When it comes to allergy and intolerance to foods, individual ingredients are always to blame. Pets are likely allergic to specific ingredients in a diet, such as beef, chicken, soy, fish, milk, corn, wheat, etc. In addition, some commercial foods contain high amounts of certain substances (e.g., histamine, saurine) that can cause problems all by themselves in susceptible individuals. However, many owners tend to blame brand names, preservatives, or a suspicious ingredient listed on the package. Many others fail to get the correct diagnosis because they refuse to believe that the problem they are seeing is diet related. They may be hesitant because they have been feeding the same diet for years without problems, or they feel that the problems are not worsened at feeding time. It's time

Self-trauma due to excessive licking, associated with allergies. Notice the hair loss over the "rump." Photo by Dr. Dunbar Gram.

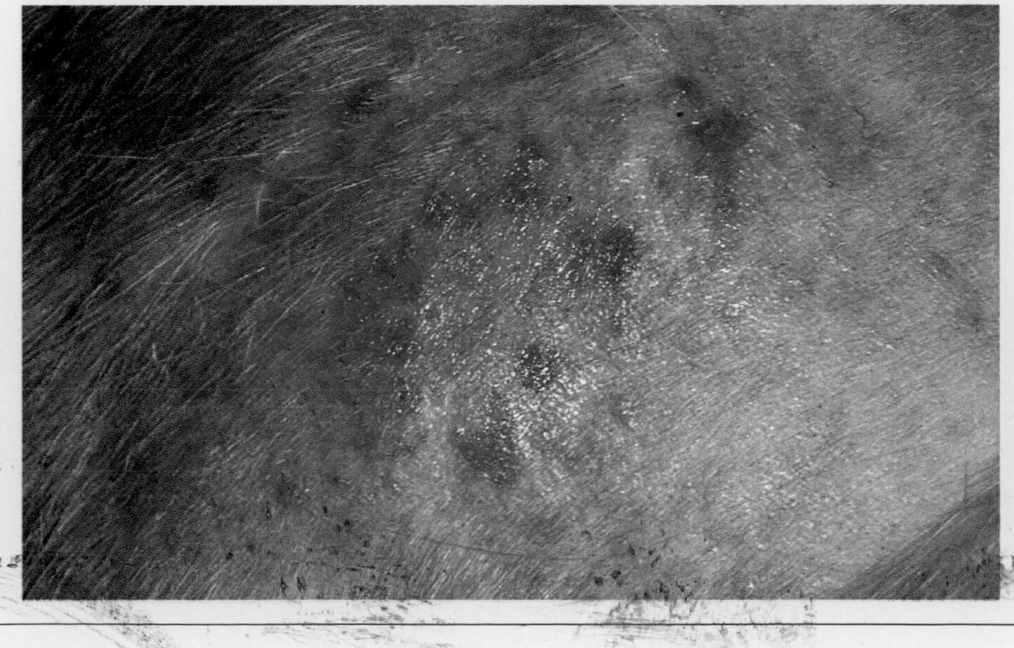

to lose all your prejudices about what you think about adverse food reactions and learn some basic truths.

How Do I Know if My Pet Has a Food Allergy/Intolerance?

It is almost impossible to confirm the diagnosis by switching from one brand of commercial food to another. Since most of the ingredients in pet foods are similar, merely changing brands or types of food is helpful only if you are lucky enough to change to a food that does not include the problem-causing ingredient.

A home-made elimination (hypoallergenic) diet is the best way to confirm a diagnosis. It must consist of ingredients to which the pet has never been exposed. This will pinpoint an adverse food reaction in most cases, but not differentiate between allergy and intolerance. A single protein source (a protein source to which the pet has not been exposed) can be combined with a carbohydrate source such as rice or potatoes and the entire diet fed for 6-12 weeks. The meat can be boiled, broiled, baked or microwaved (but not fried), and the rice should be boiled prior to serving. They are mixed at one part meat to one part rice or potatoes. The mixed ingredients should be fed in the same total volume as the pet's normal diet. During the trial, hypoallergenic foods and fresh, preferably distilled water must be fed exclusive of all else. Absolutely nothing else must be fed, including treats, snacks, vitamins, chew toys and even flavored medications. Access must also be denied to food and feces of other cats and cats in the household.

If there is substantial improvement while on the diet, further investigation is warranted. If there is no improvement, diet is not a significant part of the problem; and the diet can be discontinued. The intention is not to feed these diets indefinitely. They are not nutritionally balanced for long-term feeding. They are meant only to be fed for the 6-12 weeks necessary to determine if food is implicated in the problem. If the condition improves while on the diet, challenge feeding will be needed to determine which ingredient(s) is (are) causing the problem.

By now, most pet owners have heard about blood tests that claim to be able to diagnose food allergies. Surely this is a lot easier than having to prepare and feed a homemade diet for many weeks. That would be true if the blood tests actually lived up to their claims. Unfortunately, these tests can be very misleading since the results are often quite inaccurate, being reliable perhaps only 10% of the time. This is not surprising because not all diet-related problems are allergic (these tests don't identify intolerance), and those that are allergic are not necessarily caused by the antibodies measured in the blood tests. Therefore, blood tests should *NEVER* replace a hypoallergenic

Some cats have been known to eat houseplants, a number of which can be poisonous. A good rule of thumb is to keep houseplants out of reach of the family cat. Photo by Robert Pearcy.

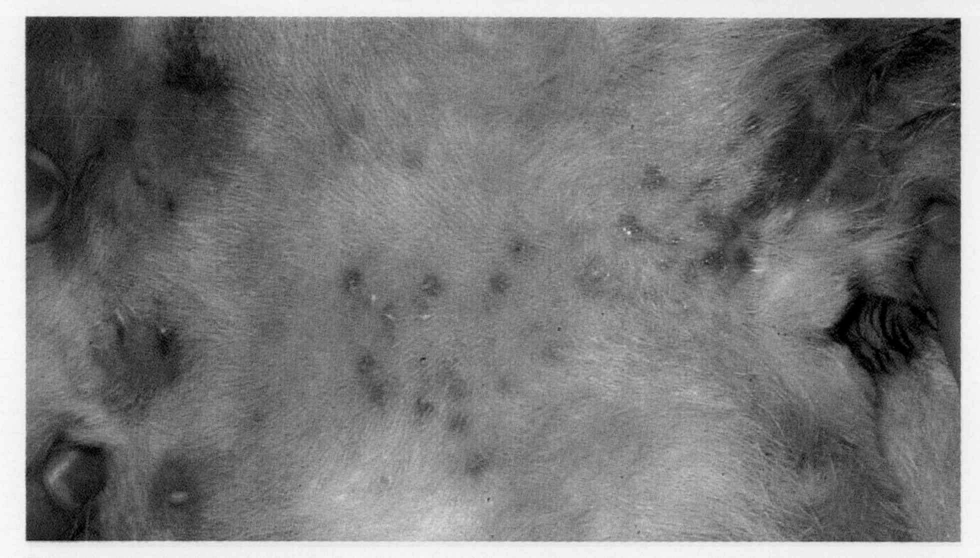

This cat did not exhibit licking in the presence of its owners but was secretively licking its abdominal area. A cat such as this is sometimes referred to as a "closet" licker. Photo courtesy of Dr. Dunbar Gram.

food trial as a screening test. More often than not, the results supplied will not prove helpful and can be misleading.

What Do I Do If My Pet Does React to His Diet?

If there is improvement by the end of the elimination diet trial, further investigation is warranted. Challenge feeding with individual ingredients should allow you to determine the dietary cause of the problem. This is accomplished by adding one new ingredient each week to the hypoallergenic diet. You then create a list of ingredients that your pet can tolerate and those that it can't. The recommendation then is to feed a balanced commercial ration that does not include the problematic ingredients. Rarely is it necessary for pets to remain on specialty (e.g., lamb-based) diets, nor is it advisable. Commercial hypoallergenic diets are suitable for owners that don't want to determine the specific cause of their pet's problem. These diets are effective about 80% of the time, for pets with documented adverse food reactions.

LINKING BEHAVIOR TO MEDICAL/ NUTRITIONAL DISORDERS

It is tempting to speculate on ways that nutritional intervention could be used to treat behavioral problems. After all, nutritional therapy could offer a relatively safe form of therapy, free from many of the adverse effects seen with drugs. Preliminary studies in humans as well as animals suggest that some conditions may indeed be amenable to nutritional intervention. Let's take a look at some examples that have been documented in pets.

Everyone recognizes that obesity in cats is to be discouraged.

Time to play. Maine Coon kitten photographed by Robert Pearcy.

The effects of hepatic lipidosis are most likely to be seen in cats that are overweight. Photo courtesy of Dr. Gary Landsberg.

However, it is not unheard of that cats may have behavioral problems while they are dieting. The cat that is an enthusiastic eater may not appreciate efforts to restrict his caloric intake. Behavioral manifestations of this outrage may include vocalizations, urine spraying and even aggression. In these cases, owners often elect to have an overweight cat rather than a behavioral basket case.

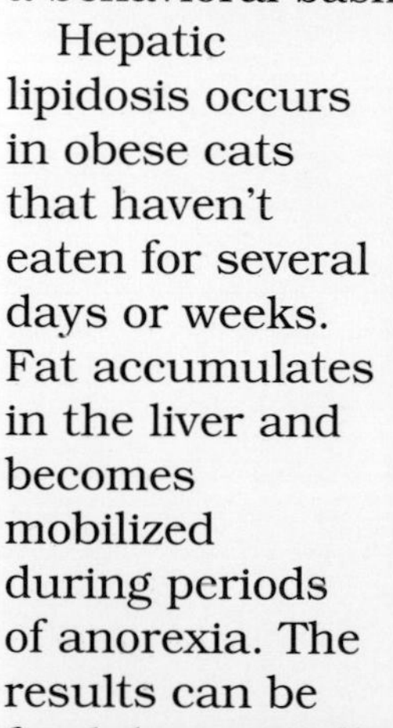

Hepatic lipidosis occurs in obese cats that haven't eaten for several days or weeks. Fat accumulates in the liver and becomes mobilized during periods of anorexia. The results can be fatal, but initially cats just appear depressed. Usually it occurs when these cats get sick and lose their appetites. Cats are very susceptible to oxidative stress, and it has been theorized that "free radical" damage is responsible for most of the ill effects. Relative deficiencies of arginine and carnitine have also been hypothesized as causative factors. Treatment is often heroic, and recovery rates may be as low as 10-20%. Tube feeding and nutritional therapy with fish oils, carnitine, arginine, thiamin and zinc have been advocated.

Taurine is an essential nutrient for cats, whereas it is non-essential for most other species. Cats can make some taurine, but not enough to meet their nutritional needs. Thus, they must be supplied with adequate amounts of taurine in their diet. Cats that receive inadequate amounts of taurine can develop eye problems (central retinal degeneration), heart disease (dilated cardiomyopathy) and reproductive problems. Recent research even suggests that taurine depletion affects immune function and hearing. All of these clinical manifestations have behavioral components. Blood levels of taurine can help establish a diagnosis. In the early stages, treatment with taurine can effectively restore health. In the past decade, most pet food manufacturers have fortified their diets with extra taurine. It is also known that cats fed a commercial canned diet have twice the taurine requirement as those fed a commmercial dry diet.

Check with your vet if you have any concerns about your cat's diet. Photo by Isabelle Francais.

Two minerals have also recently been implicated in medical problems. Potassium is an essential electrolyte, which can become depleted in association with several ailments, including kidney disease, diabetes mellitus, liver disease and lower urinary tract disease. In addition, acidifying diets can also result in

Scottish Fold kitten. Photo by Robert Pearcy.

kidney disease, diabetes mellitus, liver disease. In addition, acidifying diets can also result in potassium loss. Affected cats may be lethargic, weak or listless, but most show no signs until the depletion is life threatening. In one study, one-third of all ill cats had hypokalemia (potassium depletion). Careful attention to diet and potassium supplementation are often recommended by veterinarians. Iodine content of cat foods is now also under scrutiny. It has been theorized that the high incidence of hyperthyroidism (excessive thyroid hormone production) seen in cats may be partially due to high levels of iodine in the diet. Hyperthyroid cats are often overactive, and many suffer with heart problems. There is currently no solid link between dietary iodine and hyperthyroidism, but studies are underway.

SUMMARY

Nutrition is an important aspect of medicine, including behavioral medicine. Although the impact of nutrition on behavioral disorders is only now being explored, its importance should not be minimized. This chapter reviews some of the more common ways that nutrition and behavior cross paths and some theories that warrant future study.

ADDITIONAL READING

Ackerman, L.: *Adverse Reactions to Foods.* Journal of Veterinary Allergy and Clinical Immunology, 1993; 1(1): 18-22.

Ackerman, L.: *Enzyme Therapy in Veterinary Practice.* Advances in Nutrition, 1993; 1(3): 9-11.

Ballarini, G.: *Animal Psychodietetics.* Small Animal Practice, 1990; 31(10): 523-532.

Blackshaw, J.K.: *Management of Orally Based Problems and Aggression in Cats.* Aust Vet Practit, 1991; 21: 122-124.

Butterwick, R.F.; Wills, J.M.; Sloth, C.; Markwell: *A Study of Obese Cats on a Calorie-Controlled Weight-Reduction Program.* Vet Rec, 1994; 134(15): 372-377.

Fernstrom, J.D.: *Dietary Amino Acids and Brain Function.* Journal of the American Dietetic Association, 1994; 94(1): 71-77.

Halliwell, R.E.W.: *Comparative Aspects of Food Intolerance.* Vet. Med., 1992; September: 893-899.

Kallfelz, F.A.; Dzanis, DA: *Overnutrition: An Epidemic Problem in Pet Animal Practice?* Vet. Clin. N. Am., 1989; 19(3): 433-445.

Mugford, R.A.: *The Influence of Nutrition on Canine Behaviour.* Small Animal Practice, 1987; 28(11): 1046-1055.

Schoenthaler, SJ; Moody, JM; Pankow, LD: *Applied Nutrition and Behavior.* Applied Nutrition, 1991; 43(1): 31-39.

Wallin, MS; Rissanen, AM: *Food and Mood: Relationship between Food, Serotonin and Affective Disorders.* Acta Psychiatrica Scandinavica, 19944; 89 (Suppl. 377): 36-40.

Dr. Donal McKeown graduated from the Ontario Veterinary College and then practiced small animal medicine in Washington D.C. for 16 years. In 1974, he was appointed Associate Professor of Small Animal Surgery, Department of Clinical Studies at the Ontario Veterinary College. In 1984 he became Associate Professor of Ethology in the Department of Population Medicine. Dr. McKeown retired from the University in 1994.

Dr. McKeown has contributed to several books, written many scientific articles and lectured widely in Canada, the United States and Europe. He has received the Distinguished Teaching Award and the Teaching Excellence Award from the Ontario Veterinary College and was presented with the Veterinarian of the Year Award from the Ontario Veterinary medical Association in 1989. He is past president of the Ontario Veterinary Medical Association, vice-president of Veterinary Medical Diets Inc., president of International Bio Institute Inc., and president of the Fergus-Elora Rotary Club.

Eating and Drinking Behavior in the Cat

by Donal B. McKeown, DVM
Guelph, Ontario

INTRODUCTION

We often take eating and drinking behavior for granted, but there are several aspects of the process that are worth considering. Before we look at abnormal variations in feeding and drinking, let's first take a look at some interesting facts:

1. Cats are solitary predators of small animals, birds, reptiles and insects. Though they are carnivorous animals, they can thrive on commercial diets containing a large percentage of cooked carbohydrates. In order for starches to be digested and assimilated by the cat, they must be thoroughly cooked.
2. Cats given free access to food will eat as many as 10 to 20 small meals daily.
3. Cats require higher levels of fat and protein compared to dogs and greater amounts of taurine, arginine, cysteine and methionine. If cats are fed dog foods, they would soon develop serious nutritional diseases.
4. Cats use their teeth to kill prey and tear their food, but do not grind it.
5. Cats have evolved from desert animals and thus can efficiently conserve water. Some cats fed canned food can go several months without drinking water because most canned cat foods contain more than 70% water. Clean water should always be available for cats. It has been shown that cats will drink more water if the water is clean and moving. Cats drink by cupping the end of their tongue backwards and lap three or four times before swallowing. If the end of the tongue becomes injured, the cat cannot drink.
6. There is no evidence to support the speculation that food coloring, preservatives, high levels of carbohydrates, or protein cause any behavioral change in the cat.
7. Never give chocolate to cats as it contains oxalic acid, which prevents calcium absorption.It also contains theobromine, an alkaloid toxic to cats. Cats are very sensitive to aspirin and only tolerate very low levels.
8. The alternate pressing of the front feet or treading the breasts during nursing is usually accompanied by purring thought to stimulate milk flow. This "kneading"

behavior is occasionally observed during copulation.

APPETITE

Appetite is the drive to ingest calories and is controlled by the complex interaction of external and internal stimuli. Hunger and satiety (a feeling of being full) are controlled primarily by the hypothalamic areas of the brain, current energy levels in the body, liver metabolism, and by certain hormones. External stimuli such as the physical characteristics of the food, nutritional content of the diet (levels of fat, salt, etc.) and the previous experience of the cat would influence the palatability and contribute to the appetite drive. Unlike dogs, cats eating in groups don't stimulate an increase in the level of food consumption, nor does the sight of one cat eating cause another to eat. Cats rarely show competitive or dominance aggression related to food, and they frequently eat out of the same bowl. Therefore, using food as a competitive or a facilitative stimulus to encourage cats to eat is not helpful. Drugs such as progesterone, Valium®, cortisone, etc. can be used with the above suggestions to stimulate appetite.

DIET PREFERENCES

Cats readily become highly selective and can develop diet fixations to certain foods, particularly if they do not receive a variety of flavors when they are young. Normally, cats prefer to eat small portions of a new food for several days before switching completely, even if the new food is more palatable. In general, cats prefer a novel diet to the one they are currently eating (unless they become fixated on a specific diet). It has been shown that cats that eat the same diet for long periods may prefer the familiar diet, although it is less palatable than a new one. If the new diet is less palatable relative to the current diet, the cat will eat a small amount for several days and then stop. This behavior of eating only a small amount of a new diet is thought to have evolved as a survival technique in the wild. It is therefore wise when changing diets to gradually increase the proportions of the new diet and decrease the old, over a period of five to seven days.

Palatability of food for the cat is based on smell, mouth feel and taste, in this order. Before eating, cats spend a great deal of time sniffing and testing the temperature of their food with their sensitive nose and thus are accused of being finicky eaters. Cats are very individualistic and strongly committed to diets with particular physical characteristics. Cats can roughly be divided into two groups: those that like pureed-type foods (lickers) and those that like dry foods (chewers). The palatability of a diet for cats can usually be increased by warming the food to body temperature, adding salt, meat or moisture, and making the diet acid, sour or bitter. Adding water will not only increase

As is true for other kinds of pets, obesity in a cat can be detrimental to the animal's health. Photo by Isabelle Francais.

the odor but also change the consistency. One study found that cats like lamb, beef, horse meat, chicken and fish in this order. There are wide variations in meat preferences between cats. This variation has been speculated to be caused by genetic factors and early dietary experiences. Cats generally do not like sweets, spices, or cured meats like bacon. If you are using any of the above suggestions to encourage eating, they can gradually be changed back to the original diet once the cat starts to eat.

Starving a cat to encourage eating is rarely successful and may lead to a serious and sometimes fatal liver disease, especially in fat cats that are stressed.

ANOREXIA

Anorexia, or not eating, is most commonly caused by illnesses. Upper respiratory disease interferes with taste and thus contributes to anorexia. Cats that are under stress, such as being introduced to a new home or are being boarded, may not eat.

Anoretic cats should be checked by a veterinarian and treated for disease. If the cat is under stress, anti-anxiety drugs should be prescribed. Cats that are going to be subjected to stress could be given anti-anxiety drugs several days before and maintained for a week or so after the event. Cats that do not eat after being supplied highly palatable diets (diet preferences above) can be fed by a stomach tube or force fed. Certain anti-anxiety drugs such as benzodiazepines and progesterone may also increase appetite.

Cats that go off food for a few days, especially those that are overweight, may be prone to liver problems such as hepatic lipidosis. This is a serious, life-threatening illness, the importance of which should not be minimized. It is also important to realize that many animals that are sick and not eating may also develop other nutritional imbalances, such as potassium deficiency.

PREDATION

Cats have evolved physical and sensory characteristics that make them very efficient hunters. Stalking, catching and hunting are normal inherited behaviors, whereas killing and prey ingestion are learned. Cats must learn that the prey is food. Not learning this, cats may catch and even kill the prey and not eat it. What cats

A mixed breed cat stealthily approaching its prey. Photo by Robert Pearcy.

Illness or stress can cause anorexia in a cat. Anorexia is a medical condition that requires veterinary attention. Photo by Robert Pearcy.

perceive as prey varies considerably: some cats hunt birds at the exclusion of mice, whereas others may hunt only rats or snakes. Hunting queens will go through an elaborate series of behavior to teach their kittens prey catching, killing and eating. Kittens up to five months of age not exposed to killing prey generally do not kill or eat their prey.

feeding your cat extra food does not reduce the hunting drive.

The only way to prevent a cat from killing prey is to prevent access to the prey by keeping it on a leash or indoors. Declawing a cat to reduce prey catching does not help; most declawed cats have little or no problem catching and killing prey. Placing a bell on the cat may reduce its ability to catch prey.

Cats have evolved physical and sensory characteristics that make them very efficient hunters. Photo by Robert Pearcy.

Hunger will stimulate hunting behavior only when the cat has learned that a specific prey is food. Cats will hunt unrelated to hunger. A well-fed cat on a commercial diet that has learned to kill prey will kill as many mice as a hungry one. Therefore, even

Hunting behavior can best be prevented by selecting breeding stock that does not hunt. Removing the kittens from a hunting queen at four weeks of age will reduce the number of kittens that will kill and eat prey as adults.

GRASS AND PLANT EATING

The eating of small amounts of grass or plants is normal for cats. The reasons for cats eating large volumes of grass are unknown but may suggest a gastrointestinal irritation. Grass is a bulky, rough material that cannot be digested by cats and thus may induce vomiting or act as a laxative. The eating of grass does not suggest a nutritional deficiency.

The eating of houseplants by cats is thought to be a normal behavior similar to that of eating of grass. Plants such as rhododendron, bleeding heart, hyacinth, walnut hulls, oleander, caladium, etc. are potentially toxic for cats. A complete list of poisonous plants and trees can be obtained from your local poison control center.

Some plants, such as valerian, cat tyme, and catnip, can induce hallucinogenic-like responses in the cat. Eating the newly formed leaves of the spider plant can cause a transitory hallucinogenic response in cats without resulting in any apparent harmful consequences. Cats seem to control the amount and the frequency of eating spider plant.

Catnip stimulates the central nervous system in cats when they smell, lick, or chew it. This "marijuana-like" response of head shaking, drooling, head rubbing, rolling, glazed eye and hallucinatory behavior, last for five to ten minutes. About forty percent of cats do not have the genetic capability to respond to catnip. The response to catnip is controlled by an autosomal dominant gene, and this behavior is augmented by exposure to the plant. There is no clinical evidence that exposure to catnip causes any clinical problems.

If you happen to have a plant-eating cat, you could grow grass for your cat to eat or add a small amount of fresh vegetables to the diet. To discourage plant eating, place dangerous plants out of reach or make the plant unpleasant for the cat by attaching a small cloth bag containing a moth ball to the plant or a motion-sensitive burglar alarm, etc.

Dinnertime for this mixed breed kitten. A hearty appetite is one sign of good health in a cat. Photo by Robert Pearcy.

WOOL EATING (PICA)

Pica is the sucking, chewing and eating of things that are not generally considered food. The drive to eat foreign material is not caused by a nutritional disorder or by a gastrointestinal irritation but is due to a compulsive disorder. Sucking and eating of wool are the most common examples of pica in the cat and are observed most frequently in Siamese, Siamese crosses and occasionally in Burmese. Kittens of any breed may also show nursing behavior starting shortly after weaning and may limit their sucking behavior to their own tails, claws or feet. This obsessive compulsive behavior is inherited and begins to be expressed shortly before puberty (five to six months of age) and usually lasts throughout the life of the cat. The behavior may generalize from eating wool to the eating of all types of cloth, carpet, rubber mats, etc.

These cats are usually extremely destructive, and correcting this behavior is difficult or impossible. In some cases, the pica is not curable and can be controlled only by the continual use of drugs. One can supply the cat with old clothing made from a preferred material, and hopefully the cat will eat this at the exclusion of the good clothing. By doing this, you run a slight risk of the cat eating and swallowing enough material to cause an intestinal blockage. Some cats just suck the materials; others eat small portions. Rarely do they eat enough to cause a blockage or other intestinal disturbances.

In unusual cases, adult cats have been reported eating household objects made of wood. Both wool and wood eating are caused by environmental conflicts that induce this obsessive compulsive disorder.

The most common environmental conflicts that cats experience are those with other cats, both in and outside the house, or changes to the home environment itself. Cats that have obsessive compulsive behaviors should be kept inside to avoid conflicts with other cats. Indoor cats should be kept from seeing, smelling or hearing outdoor cats. The next most common cause of conflict is due to owners who discipline their cat or because the cat is genetically programmed to be very friendly and suffers separation anxiety when the owner leaves. Even if one cannot determine the source of the conflict, the cat should be treated for obsessive compulsive disorder anyway, with the hope that the stressor was transitory and is no longer present in the environment. Wool chewing and eating can be prevented by not breeding those cats that exhibit the behavior.

THE EFFECTS OF ANXIETY ON EATING AND DRINKING

Cats that are under stress such as being placed in a new environment or exposed to a new social situation may refuse to eat until the stressor is gone or until they have gotten used to the offending stimulus. This anxiety-induced anorexia can last for very long periods and could be a serious

This cat is a wool eater. The habit of wool eating is an inherited behavior. Photo by Robert Pearcy.

heath hazard to the cat. Certainly the most common cause of a cat not eating is due to an illness, and thus it is very important to have your anoretic cat examined by a veterinarian.

Cats suffering anxiety may frequently vomit after eating. This behavior can be confused with vomiting caused by the presence of hair balls in the stomach. Cats often lick excessively because of environmentally induced anxiety and from a variety of skin diseases. This excessive licking (psychogenic dermatitis) and the associated hair loss results in the swallowing of hair that can accumulate in the stomach. These hair balls can cause a reduced food intake and vomiting.

Your veterinarian will need to prescribe medication for hair balls. The stress-induced anorexia and vomiting require the identification and removal of the stressors and the administration of anti-anxiety drugs.

If your cat is going to be stressed by moving to a new house or be introduced to a new social group, it may be prudent to prevent the stress and help the cat adjust by giving anti-anxiety drugs several days before the event and continue for several weeks after.

The anxious cat may have to be supervised when playing with common household objects such as ribbon, string, or yarn. Photo by Robert Pearcy.

GARBAGE EATING

Cats are very careful in what they eat. Thus, garbage eating has not been observed in the cat, although food stealing may occur so that garbage may be raided if it contains appealing leftovers.

OBESITY

Obesity is one of the most common nutritional diseases of cats and affects approximately 10—to 20% of the population. Being overweight may lead to serious health problems such as cardiovascular, respiratory and liver disease, as well as diabetes mellitus, and neoplasia.

Obesity is caused by the owner giving the cat uncontrolled amounts of highly palatable commercial diets, human foods and treats. Spaying or castration does not contribute significantly to weight gain in the cat. It is true that neutered cats are slightly less active, but they seem to adjust their intake to compensate. The belief that neutering induces weight gain probably occurs from incorrectly transferring information concerning the effects of castration from one species to another. Effects of castration on behavior vary widely between species. Another reason for falsely suspecting that neutering causes obesity is the fact that cats are generally neutered at the end of the growth period (six to nine months). Owners do not decrease the amount of food or change from the high-protein, high-energy growth diet to one designed for maintenance. Thus, the weight increase does coincidentally correspond to the time of neutering.

A fifteen-year-old mixed breed cat. Photo by Robert Pearcy.

Treatment of obesity in cats should start with a thorough examination, urinalysis, and blood

A sedentary lifestyle and a high-fat diet can lead to obesity. Photo by Robert Pearcy.

chemistry to identify the presence of any concurrent disease. One should next obtain the current weight of the cat and establish a normal weight for optimal health.

To be successful, we must not only change the pet's diet but also educate the owner to change his or her behavior related to feeding the cat. The correct reducing diet should be chosen and an accurate amount of food calculated. When giving the cat fewer calories, it is advisable to divide the daily feeding into two equal meals. It is important not to base the number of calories given to the cat on the normal or desired weight, as this radical reduction in calories could cause a severe or fatal liver disease. For the same reason, it is important that the cat eats the reducing diet chosen for the weight loss program. Give the cat a conservative reduction in calories, weigh it weekly and readjust the calories to obtain not more than $\frac{1}{2}$ to 1 lb. of weight loss per month.

Failure to get compliance from the owner to carry out and maintain the weight reduction program is the most frequent reason for the failure of the program. The weight reduction program should be explained fully and written instructions given to the owner. The owner should be made aware of the serious health consequences of obesity and the length of time required to obtain the desired weight loss. Owners should be required to keep a daily record of the food intake and

Body type varies significantly among the various breeds of cat. The Siamese, shown here, is long and lean in appearance. Photo by Robert Pearcy.

Head study of a young Abyssinian. Photo by Robert Pearcy.

weekly weights of the cat. This diary and the cat should be presented to the veterinarian for reevaluation and adjustments to the program every three or four weeks.

When the desired weight is reached, an ongoing program of diet control should be in place, followed by periodic checkups to prevent a recurrence of obesity. For owners that have two or more cats and only one of which requires a special diet, it can be difficult for some owners to separate cats during feeding. For cats used to eating free choice, multiple small meals can be fed under owner supervision. Alternatively, that cat can be fed in a separate room. Another option is the Smart Bowl®, a bowl with a built-in alarm and magnetic detector. A magnetic collar is worn by the cat. An alarm sounds if the cat approaches the bowl.

SUMMARY

Although we rarely give eating and drinking a second thought, these natural functions can be an important cause of behavioral problems. Whether it's a cat preying on something in the house or chewing on a wool sweater, eating disorders must be dealt with in a comprehensive manner.

ADDITIONAL READING

Beaver B.V.: *Feline Behavior: A Guide for Veterinarians* . W.B. Saunders Co. Philadelphia, 1992, pp 171—202.

Bradshaw J .W. S: *The Behavior Of The Domestic Cat* C.A.B. International. Wallington. Oxon, UK., 1992, pp 111—139.

Bradshaw J., Thorne C.: Feeding Behaviour, in Thorne C. (ed): *The Waltham Book Of Dog and Cat Behavior,* ed 2. Pergamon Veterinary Handbook Series, 1992, pp 115-129.

Hart B. L., Hart L. A.: *Canine and Feline Behavior Therapy.* Lea & Febiger, Philadelphia, PA., 1985, pp 161-165.

Houpt K.A. Feeding and Drinking Behavior Problems, in Marder A. R., Voith V. (ed): *The Veterinary Clinics of North America*, Small Animal Practice—Advances in Companion Animal Behavior, Philadelphia, WB Saunders Co. vol. 21, no. 2, March 1991, pp 281—298.

Houpt, K. A. *Domestic Animal Behavior for Veterinarians and Animal Scientists. Iowa State University Press*, Ames, Iowa, 1991, pp 286-290.

Morris D. : Catwatching, *The Essential Guide to Cat Behavior.* Jonathan Cape, London, 1986.

Neville P. *Do Cats Need Shrinks?* Sidgwick & Jackson, London, 1990, pp 161—168.

Turner C.D. Bateson P.: *The Domestic Cat, The Biology of Its Behavior.* Cambridge University Press, 1988, pp 111—147.

Dr. Gary Landsberg is a companion animal veterinarian (dogs and cats) at the Doncaster Animal Clinic in Thornhill, Ontario. He is also extremely active in the field of pet behavior. He operates a referral consulting service for pets with behavior problems, and has presented lectures in seminars throughout North America and in Europe. He has written numerous articles, edited a number of behavior texts and videotapes and is a coauthor of a number of books, including a veterinary textbook entitled the Practitioner's Guide to Pet Behavior. *Dr. Landsberg is a past president of the Toronto Academy of Veterinary Medicine and the immediate past president of the American Veterinary Society of Animal Behavior.*

Scratching and Destructive Behavior

by Gary Landsberg, DVM
Doncaster Animal Clinic
99 Henderson Avenue
Thornhill, Ontario
L3T 2K9

INTRODUCTION

Scratching is a perfectly normal feline behavior. Although scratching does serve to shorten and condition the claws, the primary reasons that cats scratch are to mark their territory and to stretch. Cats may also threaten or play with a swipe of their paws.

For cats that live primarily outdoors, scratching is seldom a problem for the owners. Scratching is usually directed at prominent objects such as tree trunks or fence posts. Play swatting with other cats seldom leads to injuries, since cats have a fairly thick coat for protection. When play does get a little rough, most cats are pretty good at sorting things out between themselves. Occasionally, rough play or territorial fighting does lead to injuries or abscesses that would require veterinary attention.

Cats that live primarily or exclusively indoors may run into disfavor with their owners when they begin to scratch furniture, walls, or doors, or when they use their claws to climb up or hang from the drapes. Claws can also cause injuries to people when the cats are overly playful or don't like a particular type of handling or restraint. With a good understanding of cat behavior and a little bit of effort, it should be possible to prevent or avoid most clawing problems, even for those cats that live exclusively indoors.

Far too many owners are preoccupied with punishing their cats for inappropriate scratching, instead of providing an appropriate scratching area. Cats that go outdoors may be content to scratch when outside, and leave the walls and furniture intact when indoors. Cats that spend most of their time indoors, however, will usually require an area for indoor scratching, climbing, and play.

SCRATCHING POSTS AND PLAY CENTERS

Since most cats love to scratch, play, explore, and

climb, they must be provided with an appropriate area for these activities when indoors. If not, don't be surprised if you come home to objects strewn all over the floor, scratches on your furniture, and your cat playfully climbing or dangling from your drapes. Building or designing a scratching post, providing appropriate play toys, and keeping the cat away from potential problem areas are all that are needed to deal with most scratching problems.

Since cats use the post for marking and stretching, posts should be set up in prominent areas, with at least one situated close to the cat's sleeping quarters. The post should be tall enough for the cat to scratch while standing on hind legs with the forelegs extended and sturdy enough so that it does not topple when scratched. Some cats prefer a scratching post with a corner so that two sides can be scratched at once.

Special consideration should be given to the surface texture of the post. Commercial posts are often covered with tightly woven material for durability, but many cats prefer a loosely woven material where the claws can hook and tear during scratching. Carpet may be an acceptable covering, but it should be combed first to make certain that there are no tight loops. Some cats prefer sisal, a piece of

Pet shops stock a variety of scratching posts and play centers. Photo by Dr. Gary Landsberg.

material from an old chair, or even bare wood for scratching. Be certain to use a material that appeals to your cat.

A good way to get the cat to approach and use the post, is to turn the scratching area into an interesting and desirable play center. Perches to climb on, spaces to climb into, and toys mounted on ropes or springs are highly appealing to most cats. Placing a few play toys, cardboard boxes, catnip treats, or even the food bowl in the area should help to keep the cat occupied. Food rewards can also be given if the owner observes the cat scratching at its post. A product known as Pavlov's Cat™ has been designed to reward the cat automatically by dispensing food rewards each time the cat scratches. It may also be helpful to take the cat to the post, gently rub its paws along the post in a scratching motion, and give it a food reward. This technique should not be attempted, however, if it causes any fear or anxiety.

If you don't provide your cat with an appropriate area in which it can scratch its claws and play, it will select a spot on its own—which may not be to your liking. Photo by D. Barron.

PREVENTING AND CORRECTING PROBLEM SCRATCHING

Despite the best of plans and the finest of scratching posts, some cats may continue to scratch or climb in inappropriate areas. At this point, a little time, effort, and ingenuity might be necessary. A brief consultation with your veterinarian may be helpful to get back on the right track. The first thing to consider is partial confinement or "cat-proofing" your home when you are not around to supervise. If the problem occurs in a few rooms, consider making them out of bounds by closing off a few doors or by using childproofing techniques such as child locks or barricades. The cat may even have to be kept in a single room that has been effectively cat proofed, whenever the owner cannot supervise. Of course the cat's scratching post, play center, toys, and litter box should be located in this cat-proof room. If cat proofing is not possible or the cat continues to

Cats investigate as well as play on play centers. Photo by D. Barron.

use one or two pieces of furniture, you might want to consider moving the furniture, or placing a scratching post directly in front of the furniture that is being scratched. Some scratching posts are even designed to be wall mounted or hung on doors. Placing additional scratching posts in strategic areas may also be helpful for some cats. Punishment (booby traps and remote devices) may also be useful for keeping cats away from specific problem areas. As a last resort, the cat could be confined to a large crate with a litter box, toys and bedding, whenever the owner cannot supervise the cat. Crate training will effectively prevent most behavior problems, such as destructive behavior and housesoiling. Should the owner ever catch the cat when it is scratching an inappropriate object, it should immediately be interrupted (preferably using one of the remote punishment techniques described below). Keeping the cat's nails properly trimmed or using plastic nail covers are also useful techniques for some owners.

DESTRUCTIVE BEHAVIOR

Although scratching is the primary cause of household damage by cats, chewing, sucking, and overly playful or investigative behavior can also be a problem in some homes.

Investigation and Play

An important part of the development of a young animal is a healthy desire to investigate and play. These behaviors, however, can lead to damage to the household as well as injury to the kitten. Preventing or correcting these problems is quite simple, provided that the owner accepts the cat's needs to play and investigate, and provides suitable opportunities and outlets for the cat to perform these behaviors. When the cat cannot be supervised, it should be left in a cat-proof area. Any of the owner's possessions or household objects that might be clawed, pounced on, explored, or knocked flying, should be either kept out of the cat's reach or booby trapped. In addition to a play center and scratching post, the cat should be provided with play toys that can be swatted, batted, or chased. Cat toys on springs and those that are hung from doors or play centers often work well. Ping-Pong™ balls, whole walnuts, or catnip mice are often fun for cats to chase and attack. Some cats like to explore new objects, so a few empty boxes or paper bags (never plastic) will keep some cats entertained until the owner has time to play. Sometimes the best solution is to get a second cat for companionship and play. Be certain that the second cat is young, sociable and playful. Punishment techniques should be avoided except as described below.

Scratching posts come in a variety of styles and designs that will fit anyone's budget. Photo courtesy of Hagen.

Chewing

During exploration and play, kittens (and some adult cats), will chew on a variety of objects. Not only can this lead to damage or destruction of the owner's possessions, but some chewing can be dangerous to the cat. Potential targets of the cat's chewing should be kept out of the cat's reach. When this is not possible the cat may need to be confined to a cat-proof room, or the problem areas may have to be booby trapped. String and thread, electric cords, plastic bags, twist ties, and pins and needles are just a few of the objects that, all too often, cats may chew or swallow.

Another common target of chewing is houseplants. The best solution is to keep the cat away from household plants whenever the cat cannot be supervised. Booby traps may also be effective. Some cats may be interested in chewing on dog toys or biscuits, and feeding a dry cat food may help satisfy a cat's need to chew. In some cats the desire for chewing plant material can best be satisfied by providing some greens (e.g., lettuce, parsley) in the food, or by planting a small kitty herb garden for chewing.

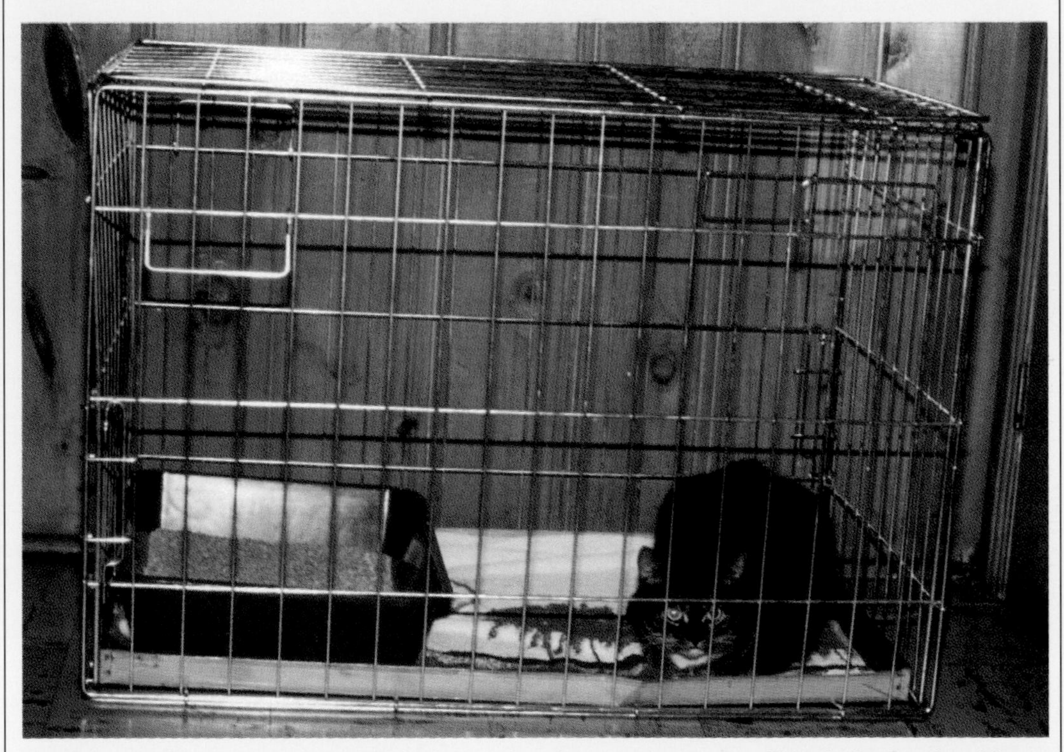

A crate can be effectively used to protect the house when the owner is not home. Photo by Dr. Gary Landsberg.

Fabric (Wool) Sucking

Although sucking on wool or other fabrics may be seen occasionally in any cat, the problem is most commonly seen in Burmese and Siamese cats. Although some cats do grow out of the problem within a few years, the problem may remain for life. The first step in correction is to try to provide alternative objects for chewing and sucking. As mentioned, some cats may be interested in one of the many chew toys or chew treats designed primarily for dogs. A well-cooked bone with some gristle and meat could be considered, provided the cat sucks and gnaws on the bone without causing it to splinter. Feeding free choice dry diets or high-fiber foods may also be helpful. Second, be certain that the cat has plenty of play periods with the owners, or even a playmate to keep it

Scratching posts can be covered with a number of different materials. Some cats prefer a carpeted surface such as shown here. Photo by Robert Pearcy.

A cat scratch feeder. When the cat scratches, food is delivered to a bowl underneath. Photo courtesy of Pavlov's Cat.

exercised and occupied. Finally, cat-proofing techniques or booby traps whenever the owner cannot supervise will likely be required.

Some cats are so persistent in their desire to suck wool that more drastic measures may be required. Providing the cat with a small amount of a product containing lanolin (such as hand cream) for licking is occasionally helpful. For some cats, it may be necessary to leave the cat with one or two woolen objects to suck on, provided no significant amounts are swallowed. It has even been suggested that a raw chicken wing a day might be tried as a last resort. However, given the prevalence of *Salmonella* in uncooked chicken, cooking or microwaving would seem prudent.

TO PUNISH OR NOT TO PUNISH

Direct Punishment

Direct punishment (i.e., where the owner punishes the pet) is seldom successful. The cat often becomes fearful or aggressive toward the owners, and, at best, all that is accomplished is that the cat learns to wait until the owners aren't watching before scratching. Under no circumstances should a cat ever be punished, unless it is caught in the act of performing the behavior. Perhaps the only

Plastic nail covers can help to prevent damage to furniture and draperies. Photo courtesy of Smart Practice.

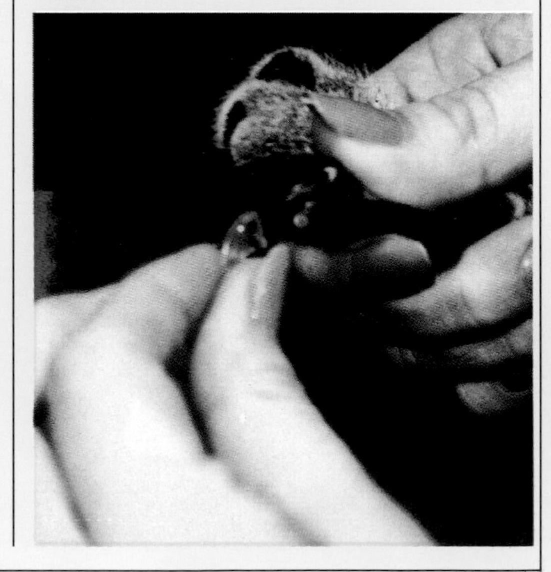

situation in which direct punishment might be successful is for the cat that swats or scratches the owners in play. Before punishment is considered however, the cat must be given ample opportunity to play. Toys that can be chased, swatted, and batted should be provided. Then, if the cat begins to swat or play attack the owners, a water sprayer, can of compressed air, cap gun or handheld alarm can be used to chase the cat away. Physical punishment should be avoided because if it is too mild, it will merely taunt the cat into more play and if it is too severe, it could injure the cat or cause fear of the owner.

Remote Punishment

Remote punishment takes a great deal of time, effort,

Above: Enticing a cat to play with its kitty play toy. Photo by Dr. Gary Landsberg. *Below:* An example of destructive chewing/sucking behavior in cats. Photo by Dr. Gary Landsberg

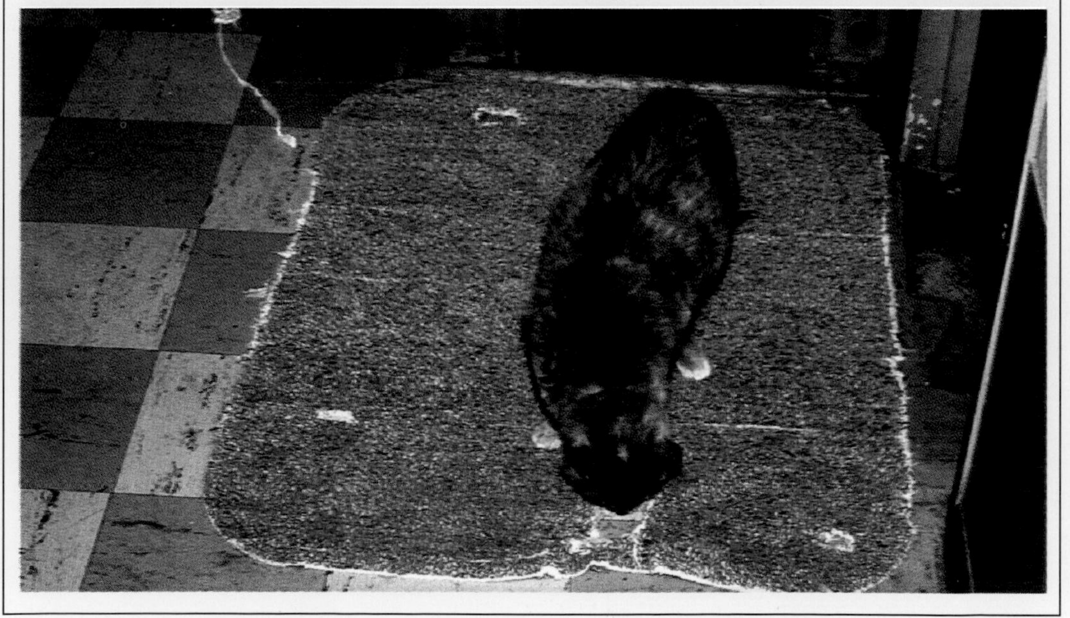

"I'm top cat in this household!" Photo by D. Barron.

practice, and ingenuity. Remote punishment techniques and booby traps are the only types of punishment that might be of use for destructiveness and climbing onto furniture. If a cat is punished during scratching or climbing while the owner remains out of sight, the cat may learn to avoid the area whether the owner is present or not.

Here's a simple example of how remote punishment might work:

- Keep a close watch on the problem area while hidden out of sight (around a corner, in a nearby closet, or behind a piece of furniture).
- As soon as the cat enters the area or begins to climb or scratch, use a long-range water rifle, noise device (such as a cap gun), or remote control device to chase the cat away.
- If the cat cannot determine where the noise or water is coming from, it should quickly learn to stay away from the area whether the owner is present or not.
- Another alternative is to set up a remote control switch near the problem area and have a device such as a Water Pik®, alarm, or hair dryer plugged in. As soon as scratching begins, the device can then be turned on by remote control to scare the cat away from the area.

For remote techniques to be

An assortment of direct training devices that can be used to help eliminate undesirable behavior. Photo by Dr. Gary Landsberg.

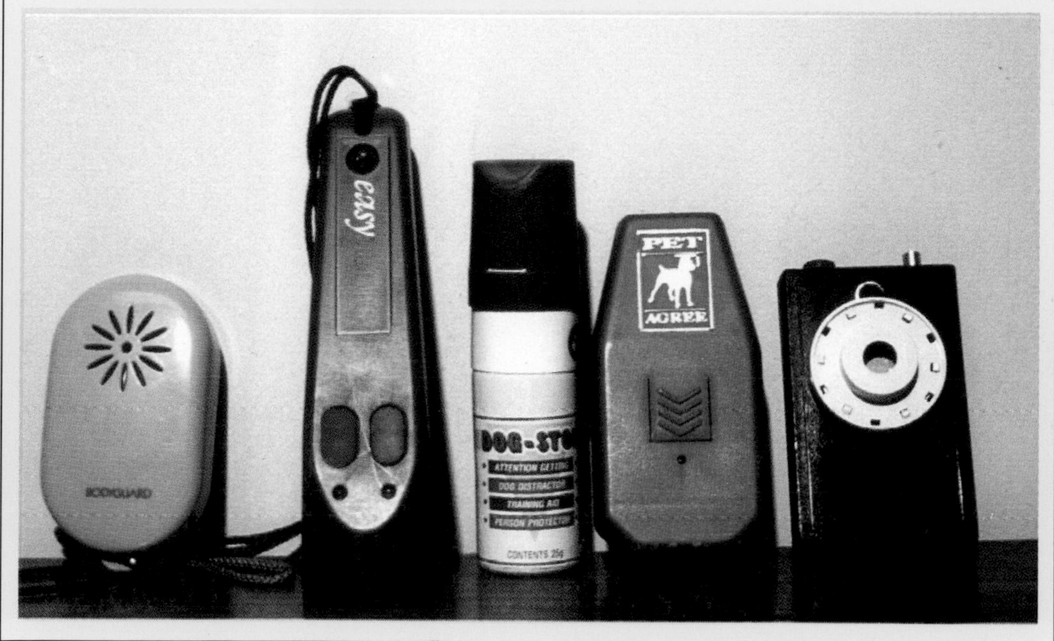

successful, there are two key elements. The first is that the owner must monitor the cat while out of sight so that the owner knows when the problem begins. A Tattle Tale™ monitor may be useful since it makes a loud beep as soon as the cat jostles the area. The second element is that the punishment must be delivered while the inappropriate behavior is occurring (while the owner remains out of sight).

Booby Traps

Perhaps the best way to discourage a cat from scratching on an inappropriate area would be to make the area less appealing (or downright unpleasant) for scratching. If the cat is scratching furniture, a large piece of material draped over the furniture may do the trick, since the cat won't be able to get its claws into the loose fabric. A small pyramid of empty tin cans or plastic containers could also be balanced on the arm of a chair so that it topples onto the cat when scratching begins. A piece of plastic carpet runner with the "nubs" facing up can be placed over a scratched piece of furniture to reduce its appeal, or a few strips of double-sided sticky tape would send most cats looking for another place to scratch (hopefully the scratching post). Mousetrap trainers, shock mats, or motion-detector alarms should do the trick when all else fails.

Most of these booby traps would also be effective for other destructive behaviors, such as chewing and sucking. Taste

Example of a remote, or hidden, punishment technique using a Super Soaker™.

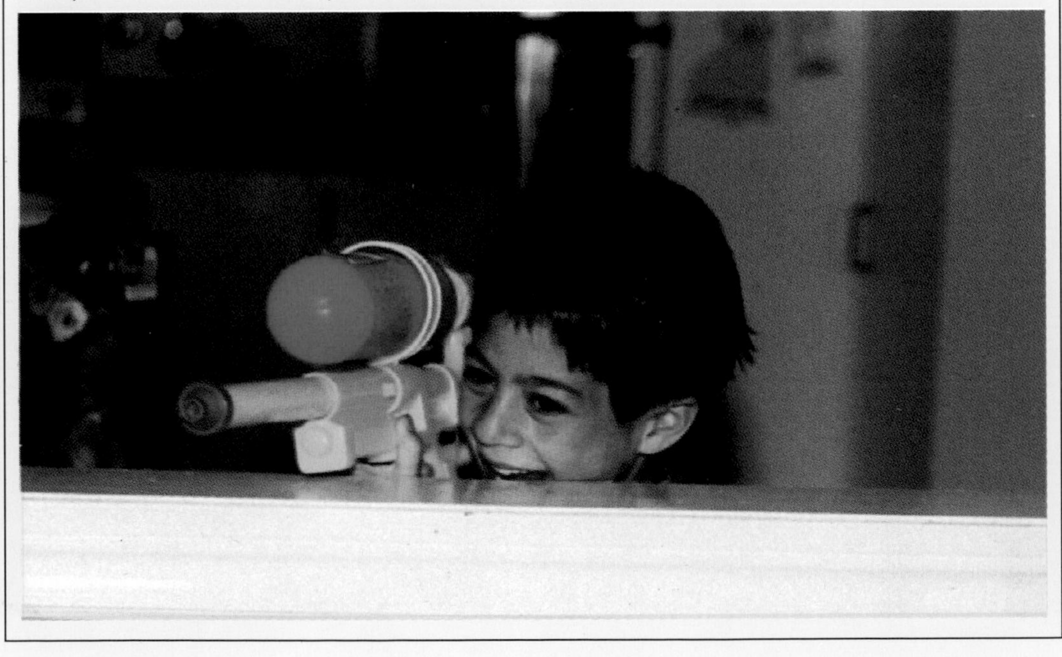

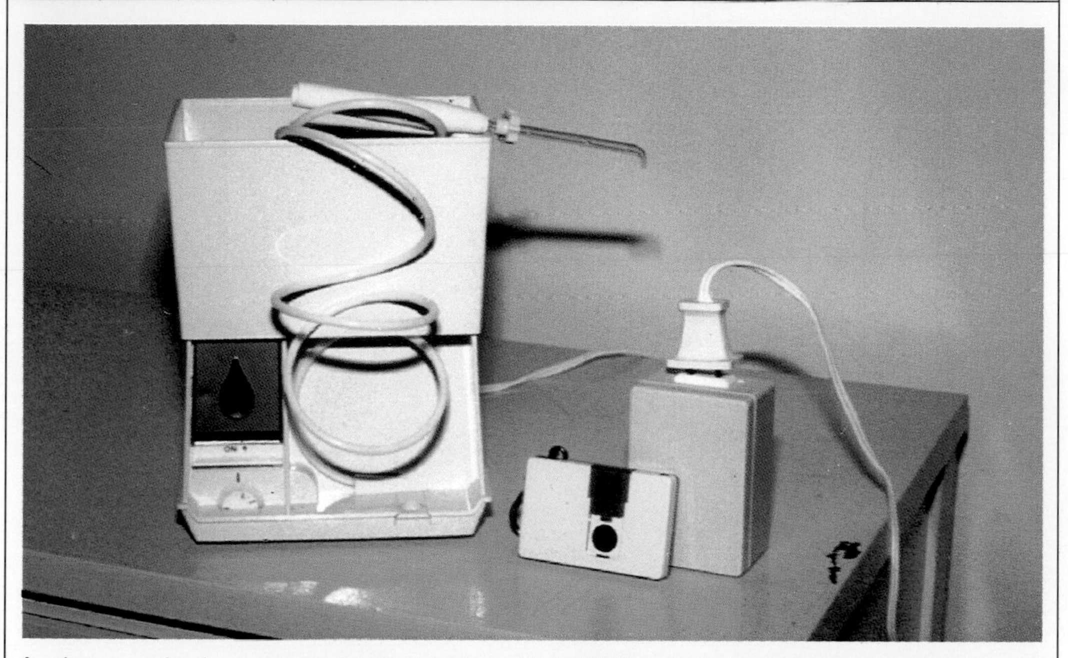

Another example of a remote punishment device: a Water Pik™ and a battery-operated remote switch and receiver. Photo by Dr. Gary Landsberg.

deterrents might also be helpful, provided they are unpleasant enough to deter the behavior. Products such as bitter apple, bitter lime or Tabasco Sauce™ are often recommended, but many cats quickly learn to accept the taste. A little water mixed with cayenne pepper, oil of eucalyptus, any non-toxic mentholated product, or one of the commercial antichew sprays often works best. To be effective, the first exposure to a product must be as repulsive as is humanely possible, so that the cat is immediately repelled whenever it smells or tastes that product again. Never leave any objects or areas untreated until the cat learns to leave the object or area alone.

DECLAWING

Declawing is a drastic, but permanent, solution for scratching problems. Certainly the behavior techniques discussed should be successful for most households. There are some homes, however, where the owners have decided that declawing is the only option if the pet is to be kept in the home. This might be the case where the cat continues to damage the furniture, or where the cat causes injuries to people during play or handling. Even the slightest scratch can have dire consequences when a member of the household suffers from a severely debilitating disease. The question therefore becomes a moral issue of whether the owners should be

A loosely draped sofa cover or a stack of precariously perched cups (unbreakable!) may deter a cat from scratching. Photo by Dr. Gary Landsberg.

able to keep their cat and have it declawed or whether the cat should be removed from the home.

There are numerous references to declawing in popular literature. Many authors write of dire behavioral and surgical complications, but these reports are based solely on myths and anecdotes. In the past few years, a number of veterinary behaviorists and pet psychologists have studied the effects of declawing on the cat, the owner, and the cat-owner relationship. In total, there have now been five scientific studies, three published in veterinary journals and two published in *Anthrozoos*, the journal of the Delta Society.

The primary reasons for declawing were property damage and the risk of injury to people or other pets. In many cases, the owners had made a concerted effort to prevent or correct the problem, but the techniques were not successful because of factors in the household, lifestyle, or cat itself. Occasionally, the health and welfare of a family member was best protected by declawing the cat (e.g., HIV patients, those undergoing chemotherapy, those with chronic illnesses such as diabetes, or when a member of the household was mentally or physically challenged). Perhaps the most startling statistic is that an estimated 50% of all declawed cats would not have been kept by their owners had they not been declawed. This means that in the province of Ontario (Canada) alone, where approximately 100,000 cats are declawed each year, as many as 50,000 cats would not have had homes if they had not been declawed.

Declawing studies have also been used to examine whether declawing causes an increase in behavior problems. In all five studies, the cat's behavior was not altered by declawing. In fact,

Above: Taste deterrents: there are a variety of products that can help to prevent destructive chewing or licking behavior. Photo by Dr. Gary Landsberg.

cats continued to scratch furniture after declawing, but caused no damage. There was no increase in behavior problems and no increase in either the severity or number of bites following declawing. Owners of declawed cats reported a higher number of good behaviors than the owners of clawed cats. Except for a few days of postsurgical discomfort, quite surprisingly, the only other owner concern was that some cats were reluctant to use the litter box when litter was replaced with paper strips. (Most veterinarians recommend that cat owners keep cats indoors and replace sandy or clay type litter with strips of paper for the first few days following

Below: With the wide array of booby traps and other training devices that are available, there is no need for a cat owner to have to "put up with" a destructive cat. Photo by Dr. Gary Landsberg.

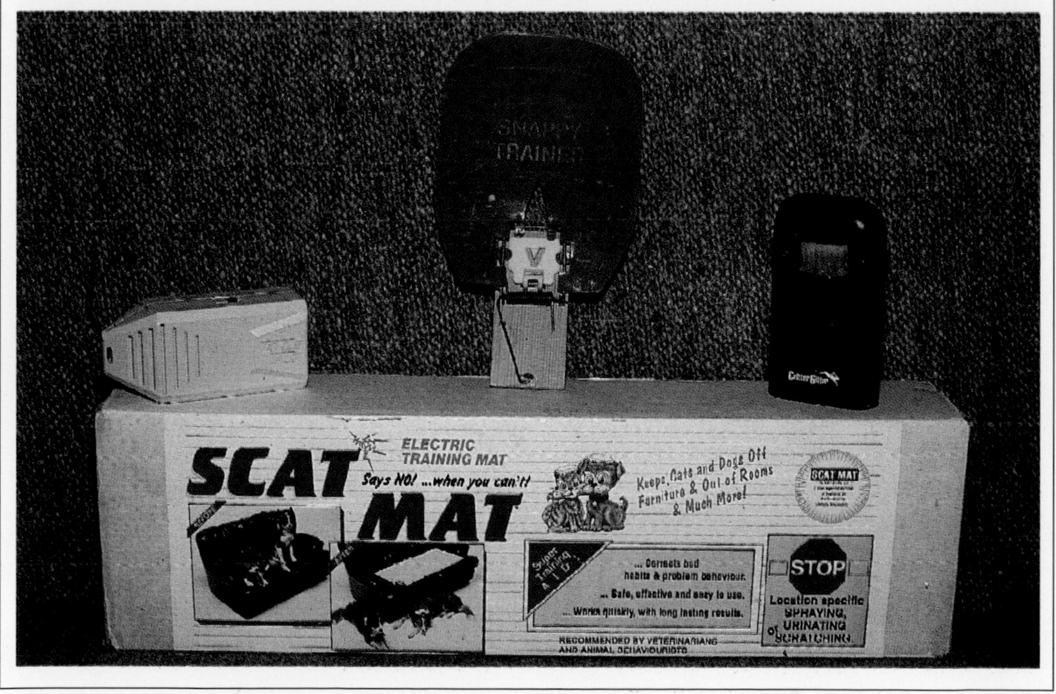

declawing). This problem has now been greatly reduced by using Yesterday's News Cat Litter™, a recycled newspaper product, for the first few days following declawing.

Finally, owners of declawed cats and veterinarians were asked to assess the effects of declawing on the cat-owner relationship. Declawing successfully met or surpassed the owner's expectations in all cases, and over 70% of cat owners made comments indicating an improvement of some aspect of the cat-owner relationship following declawing.

Once again, it is important to reiterate that scratching is a normal behavior of cats. Scratching-post training and behavior modification techniques should reduce or eliminate most forms of undesirable scratching. However, for those households in which undesirable scratching cannot be successfully corrected or prevented, declawing provides the opportunity for a larger number of families to own or keep a cat as an indoor pet.

SUMMARY

Destructive behaviors and scratching are generally considered a normal part of the cat's behavior repertoire. Rather than concentrate strictly on discouraging, punishing, or preventing these behaviors, cat owners must put some effort into providing toys, scratching posts, games, play centers and activities that can provide the cat with an acceptable outlet for its innate desires and behavioral needs.

ADDITIONAL READING

Beaver B.V. *Feline Behavior: A Guide for Veterinarians.* W. B. Saunders, Philadelphia, 1992.

Bennett M., Houpt K.A., Erb H.N., *Effects of Declawing on Feline Behavior.* Companion Animal Practice, 1988.

Bohnenkamp Gwen, *From the Cat's Point of View*, Perfect Paws, San Francisco, CA.

Bohnenkamp Gwen. *Feline Behavior Audiotapes.* James and Kenneth Publishers, Berkeley, CA.

Borchelt P.L., Voith V.L. *Aggressive Behavior in Cats.* Compend. Cont. Educ. Vet. Pract., 1987.

Bradshaw W.S. *The Behaviour of the Domestic Cat.* Redwood Press Ltd, Melksham, UK 1992.

Fisher J. (editor) *The Behaviour of Dogs and Cats.* Stanley Paul, London, 1993.

Hart B.L., Hart L.A., *Canine and Feline Behavioral Therapy*, Lea & Febiger, Philadelphia, 1985.

Hunthausen W., Landsberg G.M., *Practitioner's Guide to Pet Behavior*, AAHA Publication, Denver, 1995.

Landsberg G. Declawing revisited. Controversy over Consequences. *Veterinary Forum*, September, 1994, 94.

Landsberg G. *Cat Owners'*

Attitudes toward Declawing. Anthrozoos, 1991.

Landsberg G. *Declawing is Controversial But Still Saves Pets.* A veterinarian survey. Veterinary Forum, October 1991.

Landsberg G.M., *Feline Scratching and Destruction and the Effects of Declawing.* In: Marder, A. and Voith, V. (ed.) "Advances in Companion Animal Behavior," Veterinary Clinics of North America, W. B. Saunders Co., Philadelphia, PA, Vol 21, No. 2, Mar. 1991.

Morgan M., Houpt K.A. *Feline Behavior Problems. The Influence of Declawing.* Anthrozoos, 1989.

Morris D. *Catwatching.* Crown Publishing, NY, 1986.

Neville, P. *Do Cats Need Shrinks? Cat Behavior Explained.* Contemporary Books, Chicago, 1990.

O'Farrell V.O., Neville P. *Manual of Feline Behaviour.* British Small Animal Vet. Assoc. Kingsley House, Church Lane, Shurdington, Cheltenham, Glos, GL51 5TQ, 1994.

Turner, D. C. and Bateson, P. (ed.), *The Domestic Cat: The Biology of Its Behavior,* Cambridge University Press, New York, 1988.

Welcome Home Your New Friend: Your New Cat (video). Veterinary Marketing Services, Toronto, Ontario, Canada.

Katherine Houpt, VMD, PhD is a Professor of Physiology and the Director of the Behavior Clinic at the College of Veterinary Medicine, Cornell University. She is a charter diplomate of the American College of Veterinary Behaviorists and author of Domestic Animal Behavior *(Iowa State University Press). Her research involves the behavior and welfare of companion animals.*

Dr. Concannon teaches physiology and reproductive endocrinology at Cornell University. He conducts research on the reproductive biology of dogs and cats, with a special interest in the hormonal control of ovarian cycles and pregnancy. He is the author of over 100 publications in reproductive biology and the editor of two books on dog and cat fertility and infertility.

Sexual and Maternal Behavior in Cats

By Kathcrine A. Houpt, VMD, PhD
Diplomate, American College of Veterinary Behaviorists
and Patrick W. Concannon, PhD
Department of Physiology
College of Veterinary Medicine
Cornell University
Ithaca, NY 14853-6410

INTRODUCTION

The major problem of cats is their numbers. There are too many cats for the available homes. The result is that many cats—millions each year in the USA—are euthanized by animal shelters. One reason for the overpopulation of cats is their reproductive efficiency. Part of this efficiency is due to their sexual behavior. Female cats (queens) usually do not ovulate—release a fertile egg—until they have been bred. This phenomenon is called induced ovulation. It means that the queen does not "waste" any ovulations. Furthermore, she mates with many males (tomcats), so that if a tom fathers defective or weak kittens, another will probably father healthy kittens in the same litter. The queen will not have "wasted" her 64 days of pregnancy.

MALE SEXUAL BEHAVIOR

Male puberty occurs between 8 and 12 months of age and, although male kittens may demonstrate elements of sexual behavior in their play, true sexual behavior does not appear until after puberty. The male cat approaches the female, grasps her by the nape of her neck, mounts her and breeds her. The male first mounts fairly far forward on the queen's back so that his hind legs massage her sides. This massaging will stimulate her to arch her back and hold her tail to the side. Eventually, and while still maintaining his "neck-bite," the male will position himself so that he can penetrate her. Copulation takes place in less than 30 seconds. As the tom withdraws, the female usually shrieks. This is a reflex response to the tiny spines on the tomcat's penis. Intense and/or repeated stimulation by the penis appears to be necessary to induce ovulation. After copulation she will roll repeatedly. The queen may even swipe at the tom with her claws.

In a large group of free-living cats, several males will court the female. They remain near her without fighting among themselves, but they will repel any other male. The queen sits for a few minutes, then runs a few yards and then sits again. The males take turns mounting and mating her, and she may have a litter composed of kittens with several different fathers. There is little aggression among the males at that time, but the typical caterwauling tomcat fight occurs when a strange and presumably unrelated male invades a territory.

SPRAYING

Spraying is urine marking, not simply eliminating. The cat backs up to a vertical surface such as a wall, twitches his (or her) tail and squirts urine backwards. The urine of tomcats is strongly scented; even humans can smell it. We assume that the odor serves to mark the tomcat's territory. Tomcats usually have large territories that encompass several female territories. Because the tom can't be everywhere at once, he leaves urine marks at nose level in the areas of highest usage.

Queens, especially farm cats, also spray, not only to mark their smaller but still exclusive territories, but also to announce

Copulation. Note the neck bite of the male.

Post-copulation aggression: the female turns on the male.

their sexual status. A queen is most likely to spray when she is in heat (estrus).

The odor of an adult tomcat may serve to suppress sexual behavior in younger males, so they will either continue to act like juveniles or leave his territory for new areas where they can spray, fight and breed like adults. Unfortunately, many young males are killed while crossing roads in search of a new territory. That alone is a good reason to neuter cats.

GAPE

An interesting manifestation that is associated with sexual behavior is the gape. The cat lifts his head, curls the tip of his tongue to the roof of his mouth and stands gazing blankly for nearly a minute. This gaping usually occurs after the cat has sniffed or licked something such as the urine of another cat. The gape introduces the urine he has licked farther into his nose and allows him to perceive it with a special sense humans don't have—using a sense organ we don't have—the vomeronasal organ. The cat may be able to identify the sex, the sexual stage (in heat or not in heat; neutered or not) and be sexually stimulated by it. Male cats gape most often, but females will too.

FEMALE SEXUAL BEHAVIOR

The first estrus, or heat, in many cats occurs from seven to nine months of age, but, in

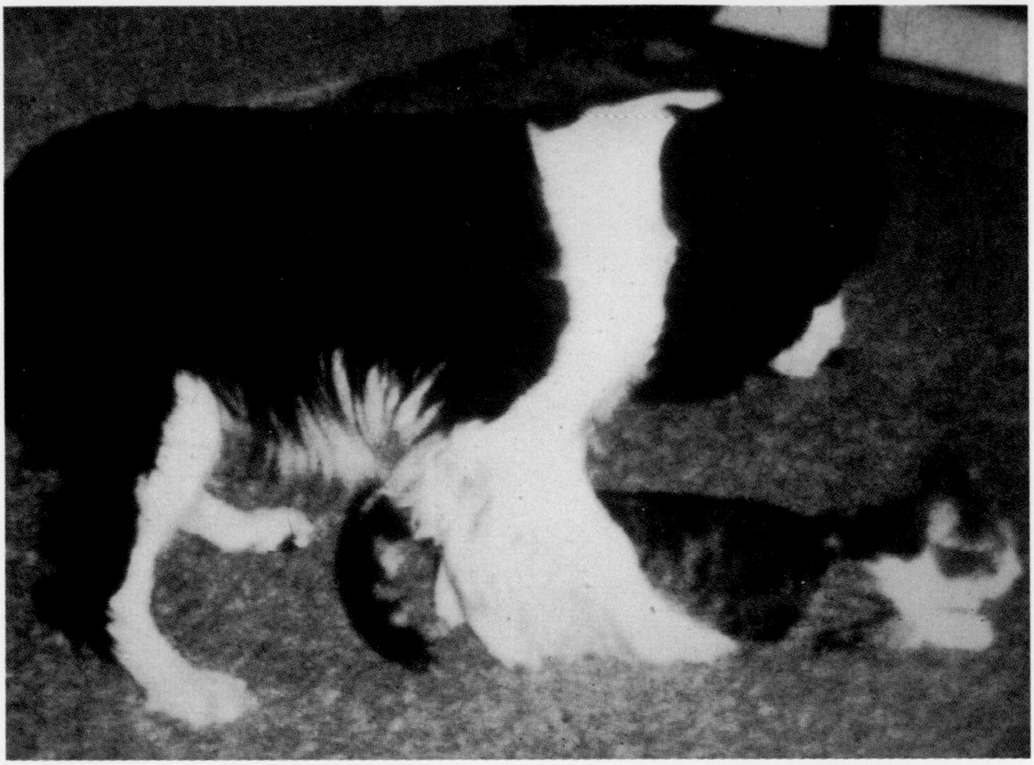

A queen in heat displaying lordosis (arched back) to a dog. Photo courtesy of Dr. Donal McKeown.

some, it may not occur until after 1 year of age. Estrus usually lasts five to ten days. After another one or two weeks, the cat will be in heat again unless she has been bred or has stimulated herself to ovulate or the season is over.

In contrast to the canine bitch who announces her sexual condition only by dripping blood on the carpet, the female cat or queen is quite blatant about her reproductive state. She is very vocal. In fact, heat in cats is termed "calling" because the loud meows are so characteristic. This is even more likely to be the case if the cat is a Siamese or other loud-voiced breed. She will rub much more often than usual, roll and sometimes spray. In addition, the cat may exhibit lordosis, the crouched position with hindquarters elevated that is the copulation position of queens. She is most likely to do this if scratched on the back. The cure for the excessive vocalization is simple—neuter the cat. In contrast to males, females rarely show sexual behavior after neutering.

Cats are seasonal breeders; they come into heat in the late winter, spring and summer. That is when the days are long. Queens will come into heat many times in a given season until they are bred. When they are bred they ovulate. This is termed "induced-ovulation" and greatly increases the chances that a breeding will be fertile. This means that cats are very efficient at breeding and explains why animal shelters are full of unwanted kittens. A female cat may mate many times a day for several days in a row—in some cases more than 50 times before she goes out of estrus. Cats may have individual preferences for males so another tom should be used if the first one is refused. Because cats are seasonal breeders, exposure to long days (12-14 hours) of light and to other cycling females will increase the likelihood that the queen will come into heat.

MATERNAL BEHAVIOR

Pregnancy lasts approximately 64 days (63-66). During pregnancy, the queen's nutritional needs increase gradually so that by the end of pregnancy she will need twice as much as she did before breeding. Pregnancy can be verified by palpation (feeling) of the kittens in the abdomen, by ultrasound or by radiograph (X-ray). The skeleton is calcified by 42 days and therefore will be visible on a radiograph at that time.

It is always difficult to predict exactly when a cat will give birth. There will be milk present in the mammary gland (breasts) one to seven days before parturition (birth). The queen will search for a nest area one to two days before parturition. A nest box lined with shredded paper or old towels should be provided in a dark, quiet place, but do not be surprised if she chooses another nest site. She probably will not be hungry the day of parturition, and her body temperature will fall. As labor begins she will meow, scratch at her bedding and pace, stopping now and then to lick herself frantically, especially her anogenital area. Delivery of the first kitten takes the longest. The queen may rest for an hour afterwards, but the whole litter should be delivered within two to six hours. If more than two hours elapse between births, the cat should be taken to the veterinarian. The average litter size is three to five kittens. After each kitten is born, the queen will lick away the fetal membranes and eat them. She will chew through the umbilical cord and lick the kitten. Once the next bout of labor begins, she will ignore the kittens and begin to pace and groom herself.

The queen should be fed free choice or three to four times her normal (nonpregnant) diet during lactation. She will stay with her kittens all of the time for the first day or so. Even later, when she does leave to eat and eliminate, she will still spend 70% of her time with them. She lies on her side encircling the kittens, thus

American Shorthair queen and kittens.
Photo by Robert Pearcy.

A recently spayed female is encouraging kittens to nurse from her rather than from their mother.

facilitating their approach to her nipples. She spends a great deal of time licking the kittens. She licks the perineal (hind) area to stimulate urination and defecation by the kittens. People who rear orphan kittens should rub the kitten under the tail with a moistened cotton swab to stimulate elimination for the first three or four weeks.

If the kitten crawls away from the nest it will cry. The cry arouses the queen, who will find the kitten, pick it up by the scruff of the neck and carry it back to the nest. The kitten hangs limply from its mother's jaws. The reaction to scruff-holding persists to some extent in adult life and can be used to restrain a fractious cat. It is also the basis of "clipnosis," in which a plastic clamp or hemostat placed on the loose skin of the back sedates the cat enough for a minor, painless procedure such as taking radiographs (X-rays).

FOSTERING

Cats will often adopt kittens

and sometimes even adopt other baby animals such as squirrels. There is a reason why this apparently strange behavior occurs in cats more frequently than in other species. Femalc cats, living on their own, on farms, for example, often share in taking care of the kittens. Two females with kittens of similar age may nest together and nurse one another's offspring. Cats who do not have kittens of their own may also assist with care of the kittens. This behavior allows the mother to leave the kittens with another cat while she hunts for food. It is important that an adult cat is present to protect the kittens from predators such as foxes or from strange tomcats. It is particularly useful to have such a helper when the kittens are being moved from one nest to another. If some of the kittens are left alone while the mother carries the kittens one by one to a new location, they are particularly vulnerable. If there is a central source of food, such as milk on a dairy farm, a queen will move her nest closer and closer to the source of food. Apparently, the dangers of predation become less and the need for nourishment greater as the kittens become older.

Maternal behavior toward another cat's kittens is reflected in the behavior of some virgin adult females and even some spayed females, who will allow other kittens to nurse. If a young kitten who has been

Self-suckling by a Siamese cat.

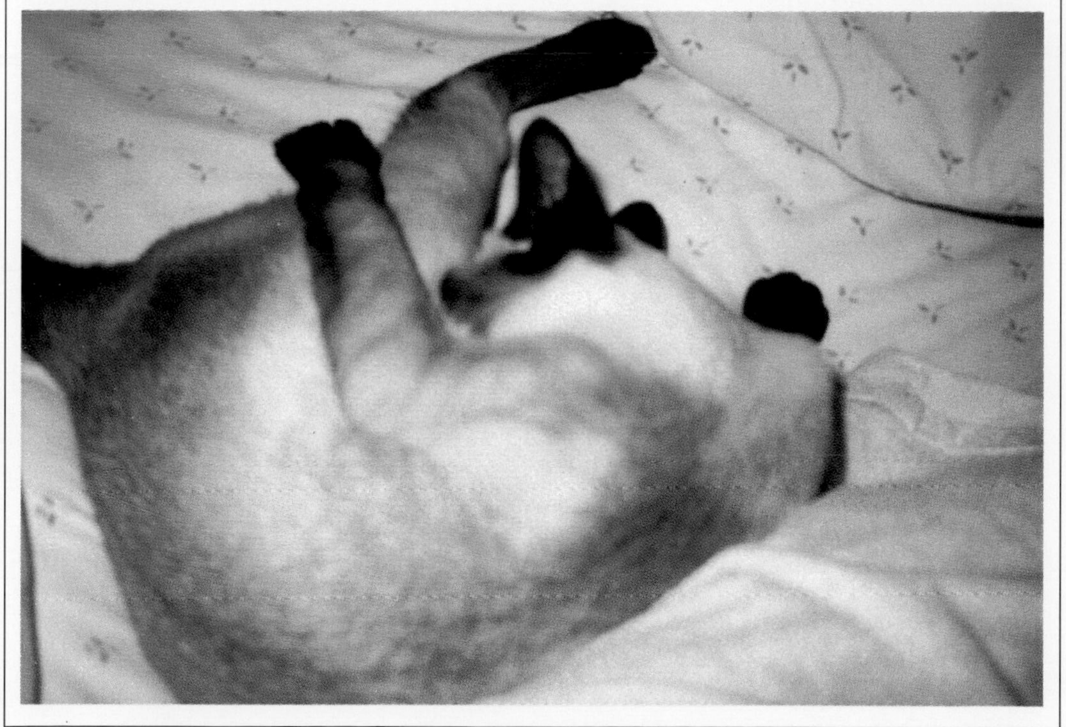

abruptly weaned is added to a household, it may be adopted by the household cat and allowed to nurse. What is even more unusual is that the "surrogate" mother may actually begin to produce milk. The females of very few other species are able to produce milk unless they have been pregnant at least once, but even virgin cats that have already been spayed can produce milk. Presumably this happens with feral cats too, so that a helper not only protects and licks the kittens but also feeds them. Although kittens and other infant animals are most likely to be accepted, even a large cat will be allowed to nurse in some circumstances.

HUNTING

A maternal behavior that is particularly interesting is teaching hunting skills. When the kittens are approximately four weeks old, the queen begins to bring prey back to the nest rather than consuming it as soon as she catches it. At first, the queen will "play" with the mouse or other quarry that she has brought back before killing it and eating it, but later she will leave the live prey for the kittens to try to catch. By ten weeks, kittens are usually proficient at hunting themselves. These hunting lessons are not essential for cats. They can learn to hunt successfully without any help from their mothers, but it does give them a head start on the skills they need for survival. The behavior of the queen with kittens may be the basis of the habit some cats have of bringing their prey home to their owners. The cat may be regarding its owner as a kitten that never seems to master the art of hunting. Most cats bring only dead prey; so either they are treating humans as young kittens, or they bring prey home because they are not hungry. Hunting and eating are separate behaviors: satiated cats will continue to hunt, and cats that normally will not hunt will do so if they are deprived of food.

WEANING

It is probably in the best interest of the queen to wean her kittens as early as possible so she can come into heat again and become pregnant, whereas it is in the best interest of the kittens to nurse as long as possible because they do not have to expend much energy or take any risks to obtain milk. This leads to weaning conflict with the queen trying to escape the kittens and the kittens trying to suckle. Some queens will wean their kittens after a few weeks, most will wean them by six to eight weeks and a few will allow their kittens to nurse indefinitely. Mothers of small (one kitten) litters are most apt to allow their kittens to nurse for many months. The amount of milk consumed by one kitten may not be much of a

Millions of cats are euthanized each year because there are not enough good homes to be found for them.

nutritional drain on the mother.

Kittens weaned too early (before six weeks) will often try to suckle from other cats, from dogs, or even on the skin or hair of their owners. Even if a kitten is obtaining all its caloric needs, it seems to crave suckling. Usually the suckling behavior gradually subsides, so the kitten is "weaned" by four months, but it persists in some cats. A few cats suckle on material rather than skin. This behavior is not to be confused with wool sucking, which is really wool chewing, and which can cause a great deal of damage to fabric. The difference is that the suckling cat usually purrs and treads with its front paws while sucking, whereas the wool-chewing cat chews material with its molars while sitting or standing with its head to one side.

If kittens are orphaned, they can be fed either by a stomach tube or with appropriately-sized bottles. If the kitten has never suckled, it cannot learn to suckle after the first few days of life. Apparently, there is a "critical period" for mastering this behavior.

NEUTERING

There is an urgent humane reason for neutering all cats. Millions of young healthy cats with no behavior problems are euthanized each year at shelters. Other cats are born and die as strays after only a few years of life, but not before they have killed some native species of birds and spread diseases such as feline immunodeficiency virus.

Cats usually do not come into heat as long as they are producing milk, i.e., lactating. This is a natural form of feline birth control. If a farm cat has a litter, the kittens should not be removed until she can be caught and spayed. A few queens come into heat shortly after the kittens are born, but this heat is usually not a fertile one.

Many people neuter their male cats although most must be urged to neuter their male dogs. The reason is that almost all tomcats spray urine in the house, whereas only a few dogs do. Some queens also spray urine while in heat. This behavior disappears in 95% of neutered female cats.

Neutered male cats almost always stop roaming and fighting, but about 10% may spray sometime in their lives. Owners can avoid spraying by keeping the number of cats in the household low (one is ideal) and avoiding stress, which leads to spraying. Neutered males may also show other elements of sexual behavior as discussed below.

BEHAVIOR PROBLEMS

Many neutered male cats continue to show some elements of sexual behavior. As mentioned above, about 10% of neutered males and 5% of spayed females will spray at some time in their lives. Another persistent sexual behavior is mounting. Neutered male cats will mount other cats, stuffed animals or articles of clothing such as socks. Most people do not object when their cat mounts inanimate objects, but will object if the cat tries to mount their arm or leg. If one cat mounts another, the mounted animal may object and a cat fight may ensue. A bite to the nape of the neck is part of male-feline behavior, but can be painful to the other cat.

Spraying is the most common behavior problem related to sexual behavior. A neutered male that sprays is usually responding either to odors of other cats or to general stress in his environment. If it is not possible to eliminate the other cats or the stress, psychoactive drugs can be used, not to sedate the cat, but to increase his ability to cope with stress.

It is rare for a queen to show poor maternal behavior, but a few cats will simply desert their kittens. Some take advantage of communal nesting to leave their kittens with another queen and leave them behind. Cats will kill their kittens if they are badly injured.

Part of maternal behavior is defense of the kittens, but this behavior is exaggerated in some queens. They will attack a dog with whom they had lived peacefully before the kittens were born, even if the dog is showing no interest in the kittens. The aggressive behavior subsides when the kittens are gone.

SUMMARY

The sexual behavior of cats has resulted in a large overpopulation of cats. Female cats do not ovulate until copulation takes place. Even after castration, male cats may show elements of sexual behavior such as spraying or mounting. Female cats show maternal behavior to kittens or other infant animals. Even spayed females may do this.

ADDITIONAL READING

Bradshaw, J. 1992. *The Behaviour of The Domestic Cat.* Wallingford, Oxford, UK.

Siegel, M. 1989. *The Cornell Books of Cats.* Villard Books, New York.

Turner, D.C. and P. Bateson. 1988. *The Domestic Cat: The Biology of Its Behavior.* Cambridge University Press, Cambridge, UK.

Dr. Leslie Larson Cooper graduated from the University of California School of Veterinary Medicine in 1980. Since that time, she has seen behavior cases as a private practitioner and as a limited-status resident at the Veterinary Medical Teaching Hospital at University of California, Davis. In her other life, she is the co-parent (with husband Phil) of sons Nick and Ben Cooper and roommate (at present) of three cats and a dog. Dr. Cooper is a Diplomate of the American College of Veterinary Behaviorists.

The Stressed-Out Cat: Dealing with Stress and Fear

By Leslie Cooper, DVM
1304 Pacific Drive
Davis, California 95616

INTRODUCTION

"Stressed out" has become a familiar phrase in recent years. Countless magazine articles now tell us how stressed we really are and what we are supposed to do about it. Humans apparently are not the only ones who get stressed: a recent article headline read, "Is Your Cat Stressed Out?" Just what is this elusive, yet all pervasive, thing called stress, and do our animals "get" it too?

DEFINING STRESS

Stress is a complex subject, and experts find it difficult to agree on a simple definition. One way to describe stress would be "the physiological and behavioral reactions of the body in response to demands placed on it." Threatening or alarming situations, even fluctuations in temperature beyond normal limits, can trigger these built-in adaptation mechanisms.

PHYSIOLOGICAL REACTIONS TO STRESS

When faced with a challenge, the body "gears up"; the heart rate increases, blood flow to the internal organs increases, and stored sugar is released into the bloodstream, ready to meet increased demands for energy. Many of these immediate effects are triggered by the release of the hormone adrenalin from the adrenal glands. The body is now ready to "fight" or "flee," depending on the circumstances. If the challenge persists, other hormones are released, among them ACTH (adrenocorticotropic hormone) from the pituitary gland. ACTH in turn causes the release of still more hormones, such as cortisone and hydrocortisone from the adrenal glands. These hormones help to continue the supply of energy-sustaining sugars. Finally, should the perceived threat continue, the previously adaptive system starts to break down. Chronic exposure to corticosteroids and other substances can cause organ systems to start degenerating, resulting in such negative effects as decreased immune response, stomach ulcers, decreased growth, and impaired reproduction. Variation in the body's response is based on the type of threat it is exposed to.

BEHAVIORAL STRESS RESPONSES

Along with the internal physiological changes, the animal's external behavior may show differences from the norm. These may be changes in routine behaviors such as eating or toilet habits, or new behavioral patterns such as hiding under the bed. Some of the more easily identified reactions are those we categorize as fearful or anxious behaviors.

Although most cat owners may detect stress in their cats by observing behavioral changes, many may miss signs of mild or short-lived stress if the behavioral changes are slight. Those behavioral changes we do see may quickly resolve themselves when the cat's body systems revert back to normal functions. Problems arise when the new or changed behaviors persist, even though the threat has ceased, or when physical illness results.

Behavioral responses to stress may involve one or more of the following categories.

Changes in Routine Habits

Unlike humans, most animals eat less when stressed. If your cat happens to be a "finicky"

Stress is something most cats would rather avoid.

The cat owner should be aware of any changes in his pet's routine behavior. Siamese photographed by Robert Pearcy.

eater, you may not notice a decrease in appetite right away. Toilet habits may also change, and in some cases failure to routinely use the litterbox can be linked to stressful changes in the environment. The cat may lose interest in playing or interacting with the owner, who may describe it as becoming "listless" or "more aloof." Diarrhea or vomiting may also be observed. Illness could cause some of these same signs, so a trip to the veterinarian, for a complete physical exam, is a necessary first step when any behavioral change becomes a concern.

Avoidance Behaviors: "Hiding Out"

Hiding is a natural way of escaping an "attack," and a cat may hide out all or part of the time to avoid contact with a visible or invisible stressor. One cat, a patient of mine, started hiding out in the bedroom closet, then gradually worked its way from closet to cupboard through the house until she settled into the kitchen cupboard right next to her food bowl. If your cat is hiding under the bed, trying to coax it out may provoke a defensive aggressive reaction, and in turn reinforce the anxiety felt by the animal.

Aggression

Cats as a species have taken the phrase "the best defense is a good offense" to heart, and are often quick to attack when frightened. Hissing, spitting, and

Waiting for the day's mail is one way to pass some time away...

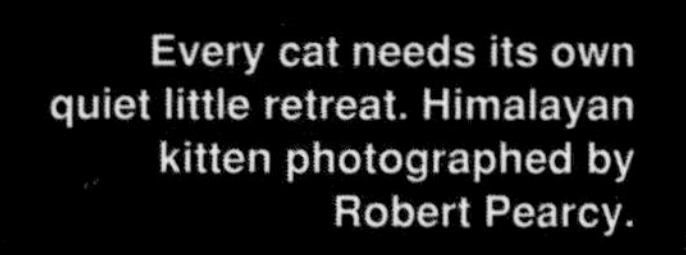

Every cat needs its own quiet little retreat. Himalayan kitten photographed by Robert Pearcy.

growling may be enough for the cat to drive off the attacker, or at least buy the cat time to beat a hasty retreat to a good hiding place. If the cat is unable or unwilling to take on the primary target, aggression may also be redirected toward innocent bystanders. Oftentimes, a cycle of aggression between two former "buddies" starts when both are hissing through the screen door at a feline intruder, then suddenly turn and attack each other. One cat may go to the fearful extreme of defensive behavior and hiding out, while the other cat takes the offensive extreme, constantly stalking and attacking its former friend. This cycle of harassment is stressful for both the harasser and the harassee, as well as for other members of the household.

During attempts to break up fights between cats, humans often become targets of redirected aggression when they try to intervene to calm them down, or as true "innocent bystanders" by simply being within striking distance. In some cats, following a triggering situation, a defensive "mood" can last for hours, or even longer. This understandably leads to a great deal of concern. Owners may even take the cat to their veterinarian, convinced that the animal has been poisoned or has some serious physical ailment.

Cats are very clean animals and devote a lot of time grooming themselves.

A contemplative kitty...

Displacement and Stereotypic Behaviors

When faced with deciding between running and hiding from an attacking animal, or standing its ground and fighting, the cat may instead choose to do something completely unrelated to the situation. This behavior is called displacement behavior; a common example is the cat who washes itself while the owner is scolding it. Stereotypic or compulsive behavior can evolve this way, as the alternate behaviors increase in frequency until they become excessive and self-reinforcing. While some compulsive behaviors seem only to bother cats' owners, others can in turn lead to related health problems. For example, excessive grooming can lead to hair loss, skin lesions, and vomiting hair balls.

Vocalization and Other Separation-Related Behaviors

Some stressed cats get "clingy." Others will simply follow their owners around crying excessively. Such separation-related behaviors are frequently caused by a permanent separation from someone, as in cases where a family member has moved out or died.

Destructive Behavior

Normal cat behaviors (scratching, playing, chewing) can be pretty destructive, but there are times when destructive behavior seems directly correlated with a stressful event. One cat I knew would jump up

Some cats have been known to climb up on counters and pull everything off of nearby shelves.

on kitchen counters and pull everything off shelves and out of cabinets whenever her owner would put her on a diet. When tying the cabinet doors shut and booby-trapping them didn't solve the problem, the cat's owner, to save her own sanity, decided to cut back on the diet.

STRESSORS

Stressors are those situations, objects or individuals in the environment that put the body on alert. Disease, extremes in weather, and finding food, water, shelter and mates can all be stressful for animals in the wild. While many of these factors are taken care of for our pets, other stressful factors may take their place. Unfamiliar surroundings, new household members, changes in schedule, and social conflicts in the household or neighborhood are all potential stressors. Some stressful stimuli are more apparent to cat than owner. For example, most cats have more sensitive hearing than our own, so humans may not even be aware of the noise causing the cat's anxiety. Other stressors, like the threat of bodily harm, seem to be shared by cat and human alike.

Social conflict is a common stressor for cats. While recent studies of cat social behavior show more flexibility and tolerance than once thought, not all cats tolerate other cats in their territories. Free-living cats have the option of leaving the area, but cats living in a household are not allowed to just pick up and leave when relationships go sour. In multiple-cat households, the effects may impact on more cats than just the ones directly involved in the conflict.

Just living with humans can be stressful, as conflicts sometimes arise when cats behave normally, but humans can't tolerate these behaviors. A good example of this is urine-marking behavior, a normal cat communicatory behavior, which can be triggered or exacerbated by stress. When cats urine mark indoors, the resulting human response (almost always a negative one) may in turn create more stress and perpetuate the problem. Inconsistent or ineffective punishment can also lead to situations where the cat hides out when the owner is around to avoid further punishment.

FACTORS INFLUENCING INDIVIDUAL RESPONSES TO STRESS

What situations prove stressful, and how intense the stress response is, depend on many factors, making each cat's response slightly different.

Early Learning

During the first few weeks of life, kittens are normally in the protected environment created by their mother. Around the third and fourth weeks after birth, the kitten's senses are operational,

The grass is always greener on the other side of the fence...

and the kitten is ready to explore the outside world. Social relationships beside the mother-kitten bond begin to develop. Cats most easily develop attachments to humans, other cats, or other species between two and seven weeks of age. While socialization can still occur later in kittens or cats having minimal social contacts during this period, it may be more difficult and result in uneasy relationships later on in life. At the other extreme, kittens who are overly socialized to humans can later show anxious behaviors when separated from their owners. As with handling, being comfortable with changes in the environment can be affected by early learning. Kittens who have little experience with exploring the environment at a young age may be more frightened of environmental changes when they're older.

Some actions that we take for granted may be a cause of stress in some animals. When you think about it, some types of petting or being picked up (other than by the scruff as a kitten) are not really natural experiences for the cat. Early handling by humans gets cats accustomed to these experiences. If a cat is not exposed to such handling, it may grow up to react with fear or aggression to something we humans think should be pleasant.

Personality Differences

Some animals just seem to be more reactive to the environment

Some cats are more social than others and enjoy sharing their home with other cats. Photo by Robert Pearcy.

Catnip is something best sampled in moderation.

than others. If the individual cat's personality is more of the shy and retiring type, the cat may have a lower tolerance for stressful situations. Several fairly recent research studies have looked at different temperament or personality types in cats, one of which is characterized as "timid" or "shy." Further study is needed to find out how big a role genetic inheritance plays in establishing a cat's basic personality.

Later Learning

Like us, cats learn from experience. If a particular experience causes feelings of fear or anxiety, then the same or similar situations later on have a good chance of causing an aversive reaction in the cat.

With all the potential stressors in the world, it's a wonder that any cat, or human being for that matter, survives and lives what could be considered a "normal" life. We must remember that a large portion of most cats' lives is spent doing relatively "non-stressful" activities: eating, sleeping, interacting positively with others. A certain amount of stress can even be helpful. Several developmental studies have demonstrated that young kittens handled for short periods of time on a daily basis showed more rapid development than their unhandled litter mates. Many behavioral responses to stress, such as fear-related behaviors, are normal, even life-saving parts of learning about the environment. A little healthy caution prevents cats from rushing into situations that could be harmful, at least until they can learn to identify the real dangers in life. Some stress is part of life, and without the body's adaptive reactions, we all would find it difficult to cope with changes in the environment.

The body's response to stress becomes a problem when it continues long enough to cause adverse effects, or when associated behavioral patterns become counterproductive in the animal's current environment.

GENERAL WAYS TO AVOID STRESS OR REDUCE THE IMPACT OF STRESSFUL SITUATIONS

Maximize Early Experiences

One way to help decrease potentially stressful situations is to expose your cat to those situations early in life. Make sure kittens are well socialized to other cats, humans, and other species they may later come into contact with. They should be frequently and gently handled, so that later handling does not come as a shock. To increase the kitten's adaptability to environmental changes in later life, early surroundings should contain lots of items for investigation and play.

Minimize Stress Caused by Change

While it is impossible (and usually not desirable) to eliminate all change from life, there are ways to help decrease the effects of stress generated by situations known to cause it. For example:

Moving

Moving to a new house or apartment is probably more stressful for cats than humans. After all, cats see just the upheaval without knowing that a move is even taking place. Try to establish some sort of routine in your new home as quickly as possible, so that your cat can count on being fed and played with at the same times each day. Unpack the cat's bed and toys first, and establish a "home corner" or "home room" where

The more things and situations a kitten is exposed to, the more adaptable it will be when it is an adult. Photo by Robert Pearcy.

the cat can stay. As the cat settles in, it may even find that empty packing boxes make great places in which to explore and play.

Introduction of new cats or people into the household

Some stressors can be difficult, if not impossible, for the animal to avoid. When the new baby comes home from the hospital, or a new roommate moves in, the cat may find it difficult to stay out of the way. Providing a "safe place", and allowing the cat to get used to the newcomer at it's own pace will help decrease stress. If possible, confine a new cat in one room when you first bring it home, and closely control the initial encounters between a new arrival and the resident cats. Expect some hissing and growling at first, and set things up so that should aggressive behavior escalate into an attack, you will be able to quickly control the situation. A gradual, unforced introduction, along with some rewarding activity such as eating or being stroked by someone they already know and accept, can further help things along.

The multiple cat household

Cats need a certain, individual amount of "space," and living together in one house decreases the living space for each cat. Providing vertical resting places (cat trees) can increase the space available. Places where cats can go to get away from each other can also help in stress reduction: an open closet door, vacant shelf, or even an empty box can be such a "hide-out."

During times of loss

As with humans, cats often need a period of "mourning" when they lose a friend or family member through separation or death. Our natural reaction to signs of distress is to try and comfort the cat. While this helps in many cases, in others we may be inadvertently reinforcing the very behavior we want to reduce by paying attention to the cat more often when it cries. Giving the animal time to adjust to the loss, reinforcing any outgoing behavior it shows, and making sure that the cat's surroundings and routine are as stable as possible may all be helpful.

WHEN A SPECIFIC STRESSOR CAUSES ANXIOUS OR FEARFUL BEHAVIOR

Fearful behavior is a normal response in environmental situations that we find threatening. When a cat repeatedly shows a fear response greatly out of proportion to what would be expected for that situation, the fear reaction is said to have become a phobia. If you have ever had a phobia, or know someone who does, you can understand just how overwhelming the emotional response can be, and how small a part logic plays in counteracting the fear. The following rules

A stressed cat can sometimes be very vocal. Photo by Robert Pearcy.

should be kept in mind when dealing with fears and phobias.

Never Force a Fear

A natural response by an owner when their cat shows fearful behavior might be to bring the animal closer to the source of the fear, presumably to "show" the cat there is nothing to fear. In doing so, they may intensify the fear by removing the cat's ability to run away from the fear-provoking stimuli, and forcing it in closer than the cat can tolerate. Another natural response is to try and comfort the cat with soothing words and petting. This usually does not reduce intensely anxious behavior, and may even inadvertently reinforce it.

Gradual Desensitization

When you can identify something specific (i.e., a noise, a person) that consistently causes the cat anxiety, the behavioral modification technique known as desensitization can be used to gradually habituate the cat. This is easiest in situations where a "gradient," or gradual, increase in the intensity of the stimulus can readily be set up. Some examples would be gradually increasing the volume of a certain type of noise, or when someone gradually decreases the distance between themselves and the cat. One of my favorite cases involved a cat who would become upset, then aggressive, whenever loud music was played, or when people spoke in loud voices. When his owners

Two cats enjoying a natural feline pastime. Photo courtesy of Dr. Leslie Larson Cooper.

A new baby in the household can be stressful to some cats. Photo courtesy of Dr. Burt Frank.

brought home their newborn baby, they knew the child's crying would trigger this response. A desensitization program using a recording of the child crying successfully habituated the cat to the point where it no longer showed agitation or aggression. Until the owners felt that a sudden noise from the baby would not provoke an attack, the cat was restrained using a harness and leash whenever the child was present.

Often food is paired with desensitization, as the internal feelings generated by a good meal tend to counteract fearful feelings in animals. This is a handy "emotional barometer": those cats too upset to eat a favored food treat generally are more fearful than they should be, and training intensity should be decreased. It may also help to provide a "safe spot" such as a cat carrier for the cat during training periods.

Adjunctive Drug Therapy

In cases where the cat spends all of its time under the bed, or when the fear response drastically interferes with normal living, the careful, short-term use of an appropriate anti-anxiety drug can be helpful. Drug therapy is used along with the behavioral modification techniques described earlier. Your veterinarian can consult

with a veterinary behaviorist for the appropiate drug and dosage. Just as the cat's response to stressors varies between individuals, so do their responses to drug therapy. More than one drug may need to be tried before the most effective one is found.

SUMMARY

While a complex and hard-to-define subject, stress can be identified by its internal and external effects. When stressed, the body prepares internally for action through a series of physiological responses and externally through changes in behavior, which may be the only indicator detectable by the cat's owners. A cat's response to life's stressors depends on several factors, including personality, early learning and later experiences. While some stress is inherent in living, its negative impacts can be reduced somewhat by a positive early exposure to stress-producing situations such as handling, and through creating a living environment that helps the cat best cope with stressful occurrences. Phobic reactions, in which the cat's response to a stressor is prolonged and out of proportion to the situation, may be reduced by careful desensitization and anti-anxiety medications when needed.

ADDITIONAL READING

Bradshaw, J. W. S.; *The Behaviour of The Domestic Cat*, CAB International, Wallingford, UK, 1992.

Dawkins, M.S.; *Animal Suffering: the Science of Animal Welfare*, London: Chapman and Hall, 1980.

Hetts, S., Estep, D.Q.; *"Is Your Cat Stressed Out?", Cat Fancy*, 1994; 37(1): 18-21.

Hunthausen, W.; *"Calming the Fearful Cat", Cat Fancy*, 1994; 37(2): 38-41.

Milani, M.M.; *The Body Language and Emotions of Cats*, New York: William Morrow and Company, Inc., 1986.

Sautter, F.J., Glover, J.A.; *Behavior, Development, and Training of the Cat: A Primer of Feline Psychology*, New York: Arco Publishing Company, Inc., 1978.

Schneck, M., Caravan, J.; *You're O.K., Your Cat's O.K.: How to Establish a Meaningful Relationship With Your Cat*, Secaucus, NJ: Chartwell Books, 1992.

Turner, D.C., Bateson, P.; *The Domestic Cat: the Biology of its Behavior*, Cambridge : Cambridge University Press, 1988.

Facing page: Excessive self-grooming (psychogenic alopecia) can be a manifestation of stress. Photo courtesy of Dr. Leslie Larson Cooper.

Dr. Amy Marder received her veterinary degree from the University of Pennsylvania School of Veterinary Medicine and later went on to complete a residency program in animal behavior there. She is a veterinary behaviorist who operates a house call practice in the Boston area, is the behavior consultant at Angell Memorial Animal Hospital in Boston, and is an instructor at the veterinary college at Tufts University School of Veterinary Medicine. She is a regular columnist for Prevention *magazine, and author of a new book,* Your Healthy Pet. *Dr. Marder is a member and past president of the American Veterinary Society of Animal Behavior.*

Introducing a New Cat to the Household

By Dr. Amy Marder
Cambridge, Massachusetts

INTRODUCTION

Even very experienced cat owners worry when the time comes to introduce their cat to a new household. Whether it's moving into a new house, in with a new roommate, getting married, or finally deciding to bring in that stray that's been hanging around forever, the process can be stressful for both human and feline. Although in most cases, cats adapt to transitions quite well, a little understanding about normal feline behavior and careful observation will make your cat's adjustment all the easier.

ACCLIMATING YOUR CAT TO ITS NEW HOME

Cats can be particularly sensitive to the sort of upheaval in their lives that comes with a

Introducing a cat to your household doesn't have to be stressful: with proper planning and preparation, it can be a pleasurable experience for all involved. Photo by Robert Pearcy.

move. Dogs, as pack animals, tend to be more attached to the other individuals in their lives and are usually fine as long as they know that their human pack members are moving also. Cats, on the other hand, are naturally more solitary and more dependent upon location. They become very attached to their territory and become stressed when removed from their home. In fact, although uncommon, there are many stories of cats returning to their old homes after a move. As most owners will move at least once during their cat's lives, here are some tips to help your feline friend cope with a move.

While you are packing before your move, try to keep the position of things in the cat's environment as constant as possible. Try not to interfere with your cat's feeding area, litter box or scratching post. Play with your cat in the packing boxes to reduce stress or put the boxes in a place where your cat can't see them. If your cat goes outdoors, make sure that your neighbors are not feeding him. If there is no food source on the old territory, it is much less likely that he will return there. By all means, make sure your cat is neutered. Neutering makes cats less likely to roam.

During the move, make sure the cat remains in a safe, confined area. Many cats escape during the chaos and confusion of a move. Friends and movers leave doors open and do not think about pets,

A kitten and a rabbit being introduced for the first time. Because they are both so young, there is a good chance that they will become friends.

Cats have been known to travel great distances to return to their old homes after a move. Photo by Robert Pearcy.

so a cage or carrier (or even boarding your cat) may be the safest place for Kitty. Also make sure that your cat has some kind of identification on him. Make your cat's carrier the last piece to move, and don't let him out until things have settled down at the new house. Then you will have the time to give him the attention he will need.

When you arrive in your new house, you might be in for surprises. After my last move, the first thing my cat Eddy did when he was finally let out of his carrier was back up to the nearest box and spray urine all over the side of it. Well, I suppose he told me what he thought of the move! Actually, although I did not appreciate his reaction, it did not surprise me. Spraying, or territorial marking, is very common after a move or when a cat enters a new territory. Both males and females spray, but it is more common in males. Some people feel that cats urine mark to increase their confidence and reduce their anxiety. If marking continues for a period of time after a move, consult your veterinarian for advice on the possible use of anti-anxiety medication.

But before you let your cat out of his box, make sure you have prepared his litter box, feeding area, bed, and scratching post. The litter box should be in a readily accessible place similar to the place where his litter used to be. If the litter used to be in the

A move to a new house can be unsettling, especially for a kitten. This little guy just wants to get away from it all! Photo by Robert Pearcy.

bathroom and your new house now has that long-awaited basement, don't put the box in the basement right away. Start by putting it in the bathroom and then move it down to the basement after your cat is accustomed to his new home. Choose a feeding area that is far away from the litter box and close to where the family spends time. The same goes for his bed and scratching post. Although you might want to, don't buy your cat a new litter box, bed or post. Keeping the things he is used to will help reduce his stress. Try to feed him at the same times he is used to. Cats are creatures of habit. Keeping to the same routine will help him cope.

If your cat will be going outdoors, try to keep him in for two weeks. This time interval is arbitrary, but it has seemed to work for me. During this time, cats come to understand that their new home is their new territory. This is where they are fed, this is where they sleep, urinate, defecate, scratch and play. It is also where they receive love and affection (if they want it!) When you do finally let your cat outside, try to supervise him at first. There probably will be other outdoor cats that have included your property within their territorial boundaries. They will not appreciate your cat invading their territory and are likely to pick a fight. After a while, they will settle their differences, but your cat may suffer a few wounds in the process. Make sure you check him every day for bites and scratches and clean them well with hydrogen peroxide. If you discover a serious or infected wound, see your veterinarian immediately.

INTRODUCING YOUR CAT TO OTHER HOUSEHOLD PETS

If your cat will be sharing his new home with other animals, take the time to make the introduction gradual. Going slowly and carefully will help to prevent disturbing, fearful, and aggressive problems from developing. Although your cat may never become close friends with the other animals, there is a very good chance that all of the animals will at least tolerate each other without fighting.

If your cat is regularly allowed outdoors and you move to a new home, keep him indoors for two weeks so that he comes to think of his new home as his own territory.

If the other animal is a cat, start by confining your cat in one room with its litter box, food, water and scratching post. At mealtime, feed the cats a distance

This is no way to make friends with the canine member of the household! Photo by Robert Pearcy.

from the door so that the cats are calm enough to eat. Gradually move the dishes closer and closer to the door until the cats do not hiss or growl at each other and can eat calmly right next to the door. Next, open the door a crack with doorstops, so that the cats can see each other, and repeat the whole process. You can also play with the cats on either side of the door. Both feeding and play will help the cats associate each other with something enjoyable and make them more likely to accept each other. During the initial introduction process, allow the cats to smell one another's beds and blankets so that they get accustomed to each other's scent.

Once the feline newcomer is using his litter box and eating normally, allow him to roam around the house while the resident cat is confined either to a room or a cage. When the hissing and growling die down, allow both of them to have the run of the house. Be sure that there are plenty of available hiding places and that you have two or more litter boxes. (Depending on the number of cats, I prefer one box per cat.) The whole process may take up to a month but is usually fine within two weeks. If at any time during the introduction process, the cats become extremely afraid or aggressive, slow down. Although mild reactions can be expected, they should not be extreme. If the cats do have a tussle, do not attempt to separate them with your bare

A kitten enjoying the security offered by its Afghan Hound companion. Photo by Robert Pearcy.

Cats can be wary of new places and new faces! Photo by Robert Pearcy.

arms. It could prove to be very painful and dangerous. Even though neither cat may ever have been aggressive to a person, a cat fight is different. Loud noises, pillows, or water are safer ways to separate fighting cats.

Dogs and cats that have had experience with members of the other species when they were young will probably adjust to each other in a short time. But those that have not had this early experience may need some extra time to get to the point where they tolerate each other. Inexperienced dogs see cats as prey and want to chase them. Cats that are not used to dogs act like prey and run or act defensively. When a dog finally accepts a cat, he is likely to relate to the cat as a member of the pack and may even protect it. But before this happens, the two need to meet each other so that a chase doesn't occur. In preparation for the introduction, brush up on the dog's obedience commands. Make sure it can sit, down, come and stay. Use food treats to keep the dog's attention, which you will need when you start the introduction.

Begin the introduction process in the same way as described for cats: feeding and playing on either side of a closed door. When they are comfortable, move to a large room and introduce them on opposite sides of the room. Keep the dog on a leash in a sit-stay or down-stay and the cat in a carrier. If the cat is loose and becomes afraid, it will run, which will elicit a chase response from the dog. Gradually move the leashed dog and the cat in the carrier closer and closer together until the dog can investigate the cat and the cat does not hiss or growl. The next step is to allow the cat to be loose in the house, but keep the dog on a leash whenever the cat is around. Finally, when the dog doesn't seem interested in chasing the cat, allow them to be free together. (Stay close by in case any problems develop).

A mixed breed longhaired kitten cautiously emerging from its hiding place. Photo by Robert Pearcy.

Two other problems of living in a household with two or more species of pets arise from the cat food and litter box. Dogs very often delight in stealing the cat's food, which tends to be more palatable than their "boring" dog food, and any tasty morsels of "doggie pate'" that can be found in the litter box. Because punishment after the fact does not change a dog's behavior (after all, the dog has already gotten his reward), the only solution is to keep both the cat's food and litter box in a place where the dog cannot get to it. I feed my cat from a special kitty counter (I have small dogs.) My sister feeds her cats on top of the refrigerator (She has large dogs.) Many other people feed and keep their cat's litter box behind a baby gate or in the laundry room or basement with a cat door or the regular door open just wide enough for the cat but not for the dog.

INTRODUCING YOUR CAT TO NEW PEOPLE

Most cats like people or at least tolerate them and do not have a problem with new people in a household. Some cats, however, have lived relatively solitary lives and were never socialized adequately when they were young. A typical scenario that I see in my practice is the single-cat owner who finally meets someone that he or she wants to share a life with, and the cat will not accept the new person. In most cases, the cat has never accepted any person and is just afraid. The same type of program that works for introducing a cat to another animal can be used for introduction to a human. At the beginning, the new person should feed the cat delicious food a distance away. Gradually the food bowl is moved closer and closer to the person, until the cat is comfortable eating out of the new person's lap. At no point should the new person attempt to pet or handle the cat; this will only make the cat more afraid. Some people want to punish the cat by spraying him with water or flicking him on his nose. Don't do it. Punishment will only make the cat more afraid and will lengthen the introduction process. Moreover, physical punishment can be dangerous, as it often makes a fearful cat become defensively aggressive. The new person should not pet the cat until he is ready and asks to be petted. A cat will start by rubbing and purring, but may still be afraid. Slowly, the cat will relax and permit petting. Very often, cats that did not grow up with children are afraid of them. The same process can be done with children.

There are many myths told about cats' and infants. My favorite is that cats suck in baby's breath and suffocate them. This myth probably got started to explain some cases of sudden infant deaths. Cats do like to sleep in cribs, however, probably because they are warm and soft. I

Facing page: Time to go out. Pet-entrance doors are convenient for cats as well as other household pets. Photo by Robert Pearcy.

Fine dining at its best! Eddie even has his own private "dining room." Photo by Dr. Amy Marder.

always recommend that people teach cats to stay out of cribs before the baby comes on the scene. There are motion alarms that can be put in cribs that will keep the cat out. A cheaper method is to use double-sided tape on a piece of cardboard that is placed in the crib.

Most cats stay away from babies when they are awake. Their noisiness and unpredictability are just too much for most cats to take. Toddlers should be watched carefully, to make sure that they don't scare or injure a new cat. If a cat is squeezed too hard and can't get away, it is likely to scratch and/or bite defensively. As the toddler gets older, he or she can take part in feeding and play—and if you're lucky, maybe even clean the litter box.

SUMMARY

Most cats adjust quite readily after a move or when they have to be introduced to new people and animals. But if your cat seems to be having some trouble, understand that it is very normal for a member of the feline species to be upset by change and displacement. Fortunately, it is also very feline to be adaptable. Give your cat time and follow the advice given in this chapter. Most likely he will do fine— but, just like a cat, at his own pace.

Facing page: Given a little time, most cats adjust to a move to a new home and will be as content and relaxed as usual. Photo by Robert Pearcy.

Robert K. Anderson is a veterinarian who has nearly 70 years of experience in modifying the behavior of animals. He has been active in research, teaching and consulting for many years at the University of Minnesota and is Professor and Director Emeritus of the Animal Behavior Clinic for the College of Veterinary Medicine. He is also Professor and Director Emeritus of the Center to Study Human-Animal Relationships and Environments, which offered the first college-level course in the nation for students to learn about relationships of animals and people. He is a Diplomate of the American College of Veterinary Behaviorists and the American College of Veterinary Preventive Medicine and has received many awards, such as the Veterinarian of the Year for the Minnesota Veterinary Medical Association and the Bustad Award as Companion Animal Veterinarian of the Year for the American Veterinary Medical Association. He now has a referral practice in Minneapolis/St. Paul, Minnesota and offers lectures and consulting services in the U.S. and other countries.

Feline Aggression Toward People

By Robert K. Anderson, DVM
Diplomate, American College of Veterinary Behaviorists
1666 Coffman, Suite 128
Falcon Heights, MN 55108

INTRODUCTION

Cats are wonderful. Studies indicate that for many people, cats are very close companions whose death would cause more grief than loss of a close relative — sometimes even more than death of a spouse. Cat people often think of cats as special guests in our homes; they are to be petted and pampered so they will allow us to give them our attention, love and affection. However, aggressive cat behavior, with biting of a person, is usually unacceptable and quickly changes previously pleasant relationships. In fact, aggressive behavior is a major reason for the disappointment, fear and even anger that breaks the human-animal bond and causes people to remove a cat from their household.

This chapter will provide two examples of aggressive cat behavior in the form of cases brought to veterinarians by concerned cat owners. Readers will also have the benefit of comments by a wise old feline named "Skycat," who will comment on some additional types of aggression at the end of the chapter.

CASE NUMBER I

History

Helen was surprised and astonished the first time that Princess, her Burmese cat, bit her hand, broke the skin and drew blood. "I couldn't believe that Princess would do something like that," she exclaimed. "I was only trying to show affection by petting and holding her in my lap. There was no warning. Just a quick hard bite and Princess scampered away before I could say or do anything," Helen told her veterinarian, Dr. James, when she and husband Tom brought Princess in for consultation.

She also showed Dr. James additional bite wounds incurred in the last several days. Helen continued by saying, "I did go to see our physician, Dr. Russell. He said cat bites are potentially

Cat lovers think that cats are wonderful and make the best of pets. Silver tabby American Shorthair photographed by Robert Pearcy.

"I want to be good!" Photo by the Barretts.

dangerous because of possible infection and he prescribed local medication and antibiotics. He also told me to see our veterinarian about treating Princess for her biting behavior or to remove her from the household."

"But I can't remove Princess," cried Helen, "I love her and she is my baby. She comes to me for attention and affection. She cuddles and lets me pet her whenever she wants petting. I think she bites only when I hold her too long. I think its my fault when she bites, but it's so unpredictable and happens so fast", said Helen.

"During the past three weeks, Helen has been trying to punish Princess," Tom told the doctor, "by smacking Princess on the nose every time she bites or acts aggressive. But Helen's timing with her hand is too slow or misdirected."

"She moves too fast for me," exclaimed Helen, with frustration in her voice. "I can't do it effectively and now she is becoming afraid to get on my lap."

"In addition, when I try to pick her up and put her in my lap, she will hiss and act aggressive. Tom agrees that punishment is not working and seems to make her more aggressive and afraid of me."

Examination

After finishing with the history, Dr. James and his staff provided general evaluation through a combination of physical, neurologic and behavioral exams. The physical and neurologic exams were within normal limits and did not appear to be related to the aggressive behavior.

Behavioral reactions during the physical, neurologic and hands-on interactive behavior exams indicated that:

a) Princess was a fairly self-confident and social cat without overt signs of fear (no hissing, eye pupils not dilated, ears not flattened against her head) when on the floor or exam table/counter. She also did not resist holding by a technician or the doctor when on the table or during taking of a blood sample.

b) Princess, when offered a food reward, responded eagerly to the moderate control of coming, sitting and being held by the doctor in his arms. But she did resist the greater control of being held in his lap and being petted over the head and shoulders by him to indicate his dominance. Princess also tried to stare down the doctor as a sign of status (top cat) by maintaining eye contact for nearly 20 seconds before averting her eyes to indicate she was being subordinate, at least for this challenge.

c) Princess did resist coming, sitting, or being held by Helen, even with a treat, and Helen could not hold Princess in her lap. Princess was less resistant to Tom; but she responded reluctantly to come and sit, and she resisted being held by him.

Comment by Skycat: Princess recognized the moderate control by the technician and doctor during the exam and did not resist them. She also, when offered a treat, eagerly responded to the moderate control of come, sit and being held in doctor's arms. However, the greater degree of resistance to the greater control of being petted over the head and shoulders, and through eye contact with the doctor, is an instinctive response by assertive cats to avoid body language that means leadership and control by a person in the mind of the cat. Also note that Princess's resistance to control was much greater with Helen and Tom. Less assertive cats would immediately respond with body language that says "I will be a subordinate cat."

In summary, Princess's behavior during all of the exams demonstrated a fairly confident, assertive and social cat without overt signs of fear. She had learned behavior to resist dominance and control by people.

Princess resisted petting except when she wished to be petted at

Facing page: Learning to understand the cat's unique personality, its moods, and instincts is one of the most interesting aspects of cat ownership. Photo by Robert Pearcy.

home. Princess also tried to show dominance with eye contact, keeping her gaze on the doctor for nearly 20 seconds on the first exam. (Eye contact is the first procedure used by many animals, including cats, to establish a pecking order, i.e., hierarchy). Cats that try to stare people down — where people move their eyes first — are indicating they want to be top cat and dominant over a person who lets them win the stare down. It is instinctive for most cats to try to be top cat with people as well as other cats.

Diagnosis

With this information, Dr. James made a diagnosis: control and status aggression by Princess toward Helen. Princess controls petting by Helen and acts as top cat with the learned aggression.

As Helen thought about and discussed the case with Dr. James, she remembered that the biting had been occurring intermittently for nearly a year, but had become a concern only after biting was severe enough to break the skin. Thus Princess had been learning to control petting by Helen, with mild bites at first, and then the control behavior, reinforced by repeated success, had escalated over time to more severe bites. It also became apparent to Helen that she was extremely permissive with Princess and allowed Princess to be "top cat" most of the time. On the other hand, Tom was not as permissive and was never bitten by Princess. Princess recognized some leadership by Tom and by the doctor and his staff at the hospital. Now Princess must recognize Helen as a leader.

Facing page: Some cats will try to establish dominance by means of eye contact, which is known as "staring down." This behavior can last up to 20 seconds or longer. Photo by Robert Pearcy.

Therapeutic Suggestions

PHARMACOLOGICAL INTERVENTION

Doctors are very careful to recommend medication only after considering the factors influencing each case including: an appropriate diagnosis, the doctor-patient relationship, the availability and efficacy of medication for this diagnosis in cats, the environment indoors and outdoors, the capabilities of the family to administer a medicine and the capability/ desire of the family to follow behavior modification programs.

In this case, the family wanted to avoid medication and seemed motivated to be successful with behavior modification for Princess and themselves.

Comment by Skycat: Doctors recognize: a) that current knowledge of appropriate medication for this type of aggression is limited, and b) that in most situations, it is best if medication is temporary (short term) to reduce the problem until therapy with behavior modification can achieve long-term success.

"We wouldn't even think of misbehaving!" Photo by the Barretts.

Behavior Modification

1. Punishment was not recommended. It usually fails for a family pet because:

   a) Some or all members of the family, such as Helen, are kind, humane, caring and do not want to hurt their pet with painful punishment.

   b) Some members of a family of all ages and strengths, are not capable of effective timing of the punishment.

   c) Punishment of aggression is potentially dangerous because it may cause the animal to respond with increased arousal and increasing aggression.

2. Positive reinforcement and humane control can help families gain dominance and leadership of their cat. This can be a very effective, humane program to help caring families change and manage behavior of their cats.

   a) Positive reinforcement of desired behavior can be carried out by all members of a family regardless of age, size and gender.

   Comment by Skycat: Cats respond eagerly to rewards of praise, petting and food, just as they learn to avoid and escape punishment and scolding. People need to help cats *accept* all members of a family as equal in leadership over the cat—in the mind of the cat, rewards may come from people of any size, age, or gender.

b) Use the giving and withholding of *attention* and intermittently food to reward cats for being eager to please people quickly. Cats must earn all rewards.

c) Avoid all scolding, punishment, and negatives such as "no ." Substitute positive commands such as come, sit, down, stay, quiet, lap, etc. to prevent or stop any unwanted behavior. Unwanted behavior automatically is prevented or stopped when a cat responds eagerly to a positive command — and as a bonus the cat is rewarded for being eager to please people.

d) Don't be permissive. Use humane control with a harness and a ten-foot indoor cord as a backup procedure to prevent failure of a cat to respond eagerly and quickly to positive commands.

3. Dominance petting and holding are important procedures in helping cats feel comfortable with all members of the family as leaders. Hold the cat on your lap parallel with your legs, in a controlled position with head pointing toward the knees. Then stroke (pet) with each hand, starting at the corners of the lips and continuing back over the head, ears, neck and shoulders to show dominance and leadership

This is "Tiger," who is about to bite. Note the aggressively held paw. Photo by the Barretts.

for two to three minutes. Do this, two or three times a day for several weeks and then intermittently several days each week.

Comment by Skycat: Cats respond to leadership of people with attentive, eager-to-please behavior and never growl, hiss, or act aggressive to any person that they accept as a leader. The key word is *accept.* When Princess accepts Helen as a leader with positive reinforcement and humane control, Princess will not try to growl, bite, or act aggressive. Instead, she will be eager to please Helen because good things happen when Princess is eager to please people.

Outcome

For five years, Helen and Tom have been bringing Princess to the clinic four to five times a year as part of Dr. James's recommended health and wellness programs. And each time they tell Dr. James how much they enjoy their relationship with Princess: "She has wonderful behavior; is so eager to please us and we owe it all to you, doctor, because you helped us become leaders that Princess accepts."

CASE NUMBER II

History

The sign on the door, in large red letters, read: "WARNING — Dangerous Cat Inside"

Dr. Erikson was not surprised and continued to ring the door bell. He was on a house call to evaluate a behavior problem of a cat named Rascal. Dr. James had referred this case to Dr. Erikson, a veterinarian, specialist in

Tiger shows his teeth while play biting. Photo by the Barretts.

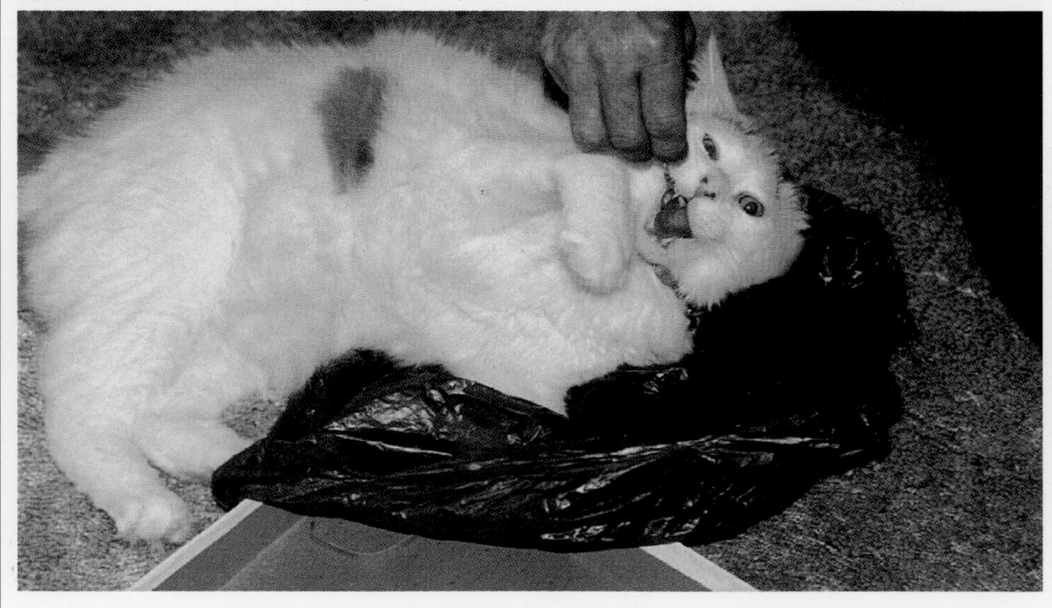

Tiger is considerably more mellow at the conclusion of his sessions. Photo by the Barretts.

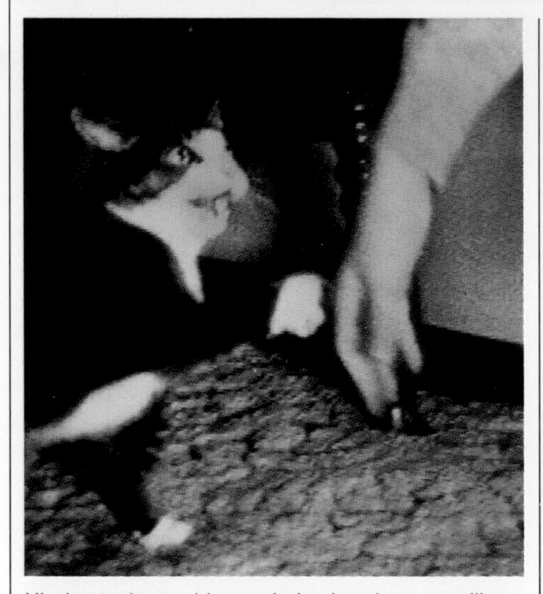

Hissing and scratching are behaviors that a cat will use when it wants to avoid control and keep people away. Photo by Anderson and Foster.

animal behavior, and a Diplomate of the American College of Veterinary Behaviorists. The cat's owner, Kathy Marek, had asked him to evaluate the cat in the cat's own environment at home, rather than in an office. Kathy's voice from inside the house door said, "Shut the screen door. I'm going to open the house door and I'm not sure I can restrain Rascal." As the door opened, Dr. Erikson heard loud hissing, which alternated with deep growling and then he saw a large white cat hissing and snarling at him through the screen. Yes, this was the right address and the sign warning of aggression was appropriate for most visitors.

To start the evaluation, Dr. Erikson asked Kathy to take Rascal away from the door and release him in the dining room or kitchen. Then Dr. Erikson entered the house. He proceeded to walk slowly toward Kathy and the dining room table, being careful to completely ignore and avoid any eye contact or recognition of Rascal to reduce the cat's concerns about visitors, particularly men. Dr. Erikson talked in a soft voice to Kathy as he sat down at the table with her and began to take the history of the problem.

Rascal was a neutered male cat about 2 years old who had been brought to her at about 12 weeks of age. He and some littermates were found living in back of an office building with no mother around. When brought to Kathy, Rascal would hiss at people, at strange sounds and at almost anything that moved.

Comment by Skycat: Hissing by a cat is typically a sign of fear. The hissing is an offensive threat that says stay away and leave me alone. Cats hiss when they want to avoid a dominant cat, a dog, a person or anything which they may perceive as a threat and be appropriate to keep away. Yes, Rascal had already learned, at 12 weeks, that he could get what he wanted: avoid control by hissing to warn/scare people and other cats to stay away. A scary defense is often a successful offense.

As he grew older and bigger, Rascal learned that he could also gain control by acting aggressively with growling, snarling and sometimes biting visitors, particularly men. Kathy usually put him away when visitors came, but Rascal had seemed to become accustomed to and accept several of her women friends. Although Kathy's boyfriend, Mike, was around the house some, Rascal still hissed when Mike moved quickly near or toward Rascal. Also, Rascal would unpredictably bite other men whenever they came into the house or moved quickly in his view.

The most recent and most severe bite had occurred when Mike moved quickly to pass by the couch where Rascal was sitting. Rascal hissed and then launched himself at Mike to severely bite and scratch a leg and then his hands as Mike bent to hit him away from the leg. As usual, Rascal then ran from the scene to avoid scolding by Kathy. This attack frightened Kathy and angered Mike. Mike said, "I'm not coming back until the cat is gone. It's either me or the cat," and stomped out of the house.

Kathy was heartbroken, but after several phone conversations and meetings at Mike's apartment, they decided to seek help from Dr. James, who referred them to Dr. Erikson.

Examination

Dr. Erikson performed the behavior exam as in Case #1. The behavioral reactions revealed that Rascal was still a very fearful cat as a result of genetic influences and the early fearful experiences in his first weeks of life.

A classic example of aggressive body language in a cat. Photo by Anderson and Foster.

Portrait of an attack cat. If your cat behaves aggressively, you should consult your veterinarian for diagnosis and therapy or referral to a behaviorist. Photo by Anderson and Foster.

Comment by Skycat: Fear is usually related to genetic temperament and then increased or decreased by experience. Experiences in the first few weeks of life—fearful or positive socialization—are a greater influence than at any other time in life.

Rascal was so fearful that he would not come to Dr. Erikson. even for a very special food reward, and the hissing was accompanied by fearful, aggressive body language (ears flattened against the head, hair erect on back of neck, body in a defensive crouch, eyes with dilated pupils, lips drawn back and mouth open in a hiss). To deal with the fear, Rascal had learned that hissing and an aggressive offense is a good defense to avoid control and have people stay away from him.

Comment by Skycat: Most aggressive cats are really fearful cats that have learned that it pays off to act aggressively to select people. In this case, Rascal was used to living with a woman, who made good things happen when she was around. He had become more accepting of women visitors. But, he was still very aggressive to men.

Diagnosis

With information from the history, examination and observations, Dr. Erikson made a diagnosis of fearful aggression toward select people, particularly men. Dr. Erikson told Kathy that the basic fear behind the aggression could not be "cured," but it may be reasonable to reduce the aggressive actions toward male visitors to a tolerable level—similar to the level Rascal exhibits with Kathy and her female friends.

To accomplish this, Mike and other men would need to cooperate to reduce Rascal's anxiety through desensitizing procedures. They would also have to provide a comfortable level of companionship with Rascal and have him eager to please people. Mike and Kathy agreed that they wanted to try therapy for Rascal and would cooperate.

Seal point Siamese. In general, members of this breed are affectionate and sociable in nature. Photo by Robert Pearcy.

Therapy

Pharmacological Intervention

Treatments recommended by Dr. Erikson included the use of an anti-anxiety medication, which would be given for about two months to reduce Rascal's anxiety so that Kathy and Mike could proceed with the behavior modification to gain leadership. Dr. Erikson referred them back to Dr. James, as their regular veterinarian, to monitor the physical and biochemical well-being of the cat while taking the medication.

Medication for behavior change must always be carefully prescribed and well monitored by the cat's veterinarian.

Behavior Modification

1. The same program as in Case #1: a) to gain leadership through rewards for being eager to please people, b) humane control with a harness and ten-foot lead to control Rascal, to prevent aggressive behavior and to ensure that Rascal never fails to respond to people and c) dominance petting and holding with intermittent rewards to reinforce leadership of people.

2. A program to desensitize Rascal to visitors, particularly men, who are welcome into the house. This program requires very careful monitoring to avoid increasing the anxiety of Rascal while gradually bringing visitors and Rascal closer together while good things happen. Soon Rascal will want visitors to come into the house so good things will happen. This program starts as soon as Rascal is eager to please people he knows. Then he can become eager to respond to visitors, particularly men.

The goal is to have a cat that is comfortable and eager to please people because good things happen when it responds eagerly to family and visitors that are welcome.

Dr. Erikson followed the progress of behavior modification, giving advice to overcome any problems that might appear during the interactions. When Kathy and Mike were doing well and Rascal was responding eagerly even to visitors, Dr. Erikson recommended reducing and then stopping the anti-anxiety medication as no longer needed to aid in reducing anxiety and fear.

Outcome (after six months)

Kathy and Mike were both happy. "You saved our relationship," they said. "We were in danger of breaking up because of Rascal, and now we feel this working together has made us closer than ever." "Thank you," they told Dr. Erikson as they looked lovingly at each other while petting Rascal lying on Mike's lap.

Comment by Skycat: People are very grateful to learn that there is something they can do to prevent or stop unwanted behavior. And it's good for the quality of life for cats also. Remember to consult your veterinarian before you get a new pet and, during the life of your pet, to make desired behavior a part of your pet's health and wellness program.

OTHER CAUSES OF AGGRESSIVE BEHAVIOR IN CATS

The two cases described in this chapter are examples of only two types of aggressive behavior — petting (status) aggression and fear-based aggression. There are many causes of feline aggressive behavior toward people, just as there are many causes of infection, vomiting, diarrhea, skin diseases, many types of cancer and other conditions affecting the health and welfare of our pets. That is why we need help from a veterinarian or a behaviorist to make a diagnosis of why our cat is acting aggressively. It could be related to the following: pain from a physical problem, maternal behavior, redirected aggression, irritable aggression, petting aggression, fear aggression, play aggression, predatory aggression, neurologic conditions, or environmental and human stimuli.

Comment by Skycat: Cats are wonderful, but you need to seek help for a diagnosis of why your cat is acting aggressively. If your cat were vomiting, you would seek an examination and a diagnosis from your veterinarian to learn why the cat is doing this unwanted behavior — vomiting. For *any* unwanted behavior, it is important to have a diagnosis of why the cat is exhibiting this behavior. This is particularly true for aggressive behavior because of the danger to people. For any aggressive behavior, please consult your veterinarian for diagnosis and therapy or referral to a behaviorist.

ADDITIONAL READING

Anderson, R. K., Foster, R. E., *Alpha-M Behavior Management System for Cats*, 1994, Alpha-M Inc. 511 Eleventh Ave. So., Minneapolis, MN, 55415.

Anderson, R.K., Hart B.L., Hart L.A., *The Pet Connection, Its Influence on Our Health and Quality of Life,* 1984, Center to Study Human-Animal Relationships and Environments, University of Minnesota, Box 734 Mayo Bldg., 420 Delaware St. S.E., Minneapolis, MN 55455.

Beaver, Bonnie V., *Feline Behavior, A Guide for Veterinarians,* 1992, W. B. Saunders Co., Philadelphia, PA.

Hunthausen, Wayne, Landsberg, Gary, *Practioner's Guide to Pet Behavior,* 1995, American Animal Hospital Devner, CO.

Neville, Peter, *Do Cats Need Shrinks?, Cat Behavior Explained*, 1991, Contemporary Books Inc., 180 N.Michigan Ave., Chicago, IL, 60601.

In veterinary practice since 1984 (graduate of the University of Montreal), Dr. Stefanie Schwartz also has an undergraduate degree in psychobiology (McGill University, 1979) and a graduate degree in ethology (University of Montreal, 1988). As a lecturer and consultant in pet behavior problems for over 10 years, she is the author of several texts on animal behavior in veterinary practice and contributes articles on small animal behavior to scientific journals, magazines and newspapers. Based in Newton, Massachusetts, Dr. Schwartz is Clinical Assistant Professor at Tufts University's School of Veterinary Medicine.

Aggression in Cats

By Stefanie Schwartz, BSc, DMV, MSc
Veterinarian, Pet Behavior Consultant
163 Lexington Street
Newton, MA 02166-1333

INTRODUCTION

Aggression is among the most vital of survival mechanisms for many species. There are many types of aggressive behaviors, and each may be expressed in varying intensity by a given individual. ``Aggressive behavior can have defensive or offensive goals. It can be directed against individuals of another species, for example, in defending against a predator or in hunting for food. Aggression can occur between members of the same species. This is common between rivals battling for a mate or in competition over territorial rights. An animal can show irritable aggression because of hunger or individual intolerance to human handling or to the discomfort of a physical injury or disease. Aggressive behavior can result from fear, for instance, when a cornered and aroused animal strikes against anything or anyone that blocks its escape to safety.

The potential for aggressiveness exists in all pet cats, whether they are confined as housepets or permitted to roam outdoors. More often than not, aggression is inhibited and is seen as abbreviated and ritualized threat displays. The risk of injury from actual fights between cats must be weighed against the intended goal of the conflict. If the overwhelming motivation of one or both cats is to advance against its opponent, a fight can occur. Still, the majority of encounters begin and end with ritualized forms of aggression, including hissing and growling, that allow rivals to assess each other while avoiding outright battles.

Aggression in cats occurs for many reasons. The context in which aggressive behavior occurs, the characteristics of the aggressor and the target or victim of the attack must all be considered in any attempt to interpret its purpose. To understand the dynamics of the situation and the goal of each individual, the current age, physical condition, sexual motivation, and past experience of the individuals involved in the conflict should be clarified. In this chapter, we will review some of the most common types of aggression in pet cats and some of the simple steps that may help to remedy them.

TERRITORIAL AGGRESSION IN THE CAT

The territorial nature of the domestic cat is a reflection of its individual predisposition, its social interactions with other cats and its wild ancestry. The use of territory in cats is three-dimensional. The cat's agility in jumping and climbing enable it to use elevated surfaces in addition to ground level. Some cats tend to range over very large areas while others seem content with a more restricted use of the space available to them. The same territory may be shared by many individuals that patrol its periphery at different times of day. This is true for housecats as well as for free-roaming pet cats.

Territorial tension between individuals is related to food availability, population density, early socialization, sex and genetic factors. Some individuals tolerate living with many other cats while others are best kept as solitary pets regardless of how they were socialized as kittens.

Territorial conflicts can occur between outdoor cats in the neighborhood or indoor cats in your home. Each cat evaluates the other from a distance. They must process as much information as possible in order to make the decision to approach or to take a detour and avoid closer contact. If individuals recognize each other from previous encounters, their past experiences and current context will influence what happens next. Meetings between intact tomcats during mating

The cat that is being held is exhibiting offensive aggression while the cat on the examination table is exhibiting fear. Photo by Dr. Gary Landsberg.

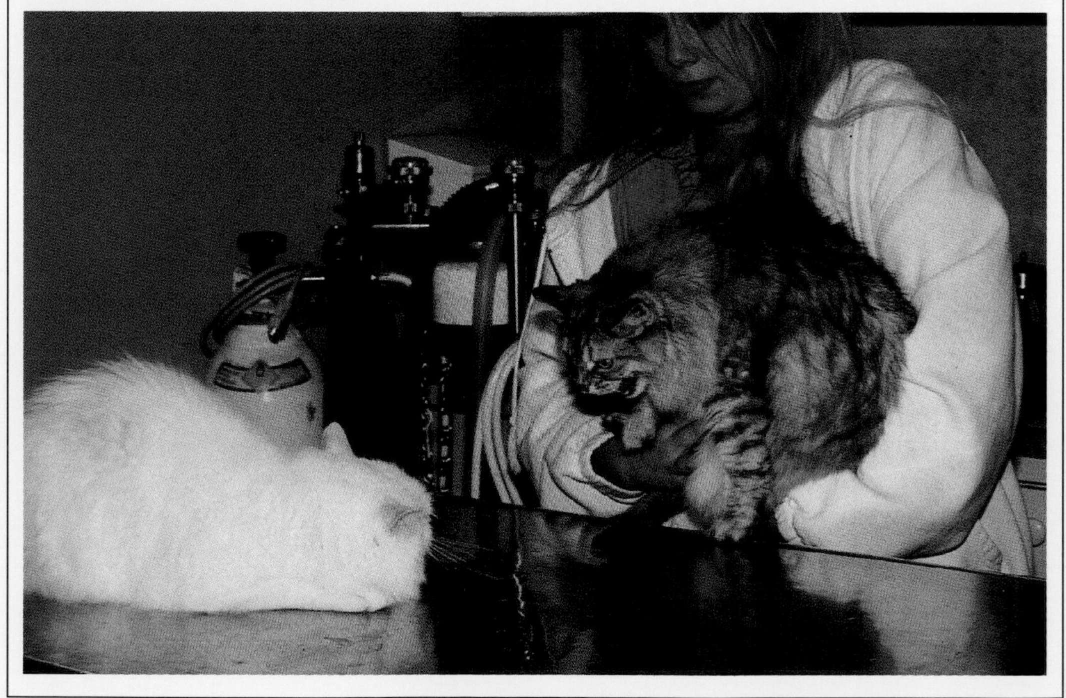

Close-up of the offensive posture by the same cat shown on the opposite page. Photo by Dr. Gary Landsberg.

season, for example, will be more likely to end in a catfight than at other times of the year, when reproductive imperatives are not as high. If neither cat is interested in either a withdrawal or a fight, they may pass each other by without further conflict. If one or both cats are motivated to a challenge, a fight is more likely. Depending on the outcome, subsequent strategies might include modifying the patrol route; rescheduling territorial patrols to different times of day; or, abandoning the territory altogether.

Conflicts between housecats can erupt from the moment of introduction or can appear following a period of calm coexistence. The age of each housecat at the time of introduction is an important consideration. Upon reaching physical and behavioral maturity, for example, kittens may awaken to their own sense of territory. They may suddenly be resented by other resident cats that formerly tolerated the presence of a juvenile. As with outdoor cats, territorial priority between housecats may be determined by subtle challenges over a period of time. Fighting between housemates is possible even between housemates that previously have appeared to get along. The more cats there are in a household, the more chances there will be disputes over territory and the use of favorite perches, food, water and even your attention.

To resolve territorial aggression between cats, the factors that contribute to it must be identified and controlled. In some cases, conflicts are brief and mild. Unless pets are getting injured, it might be wise not to interfere. If bouts are increasingly frequent and intense, however, preventive measures are recommended. It is almost always helpful to temporarily, at least, isolate the principle antagonists, confining them to their own separate quarters. The transition period may be eased with a short course of sedatives given in low doses. Medications prescribed by your veterinarian or by a veterinarian specializing in pet behavior can accelerate your cats' adjustment. With sufficient patience, time, and positive reinforcement, most cats eventually adjust to coexisting peacefully.

In severe cases, it may be necessary to permanently seclude each cat in a distinct part of your home so that they never meet. Permanent and complete separation and confinement is usually not necessary if the owner can identify specific problem situations. If conflicts tend to occur at certain times of day, for example, confine one of the cats during this period and alternate their freedom. Some conflicts tend to occur in specific areas of your home, for instance, where one cat waits in ambush. These trouble spots can be eliminated simply by blocking access to them.

Sexual tension frequently leads to fighting between cats of the same or opposite sex. Territorial outbreaks between intact tomcats or queens used for stud are easily controlled by cage confinement or separate housing. Younger cats tend to patrol wider areas than do aging animals and likely instigate more frequent fights. Males tend to roam over greater areas than do females. Neutered males may patrol smaller areas while neutered females tend to expand their territories. Whether they are permitted to go outside or not, any cat not intended for breeding purposes should be neutered. Fighting in castrated animals is less frequent and less intense, particularly in neutered male cats, compared to intact tomcats. Although territorial aggression is seen in females and males that are neutered or sexually intact, neutering will eliminate at least one of the factors encouraging more frequent and intense catfights.

When fighting between housecats becomes too severe, some pet owners are tempted to begin allowing them to roam outdoors. The multitude and magnitude of the dangers to which cats are likely to be exposed outside far exceed the danger of conflict with a housemate. A quick and easy settlement of a pet

Facing page: A tomcat defending its territory. Males tend to roam over greater areas than do females.

owner's immediate concerns cannot justify the risk to pets. As discussed, separation of the antagonists to different areas of the house is much kinder than exposing them to the infinite dangers outside. The simplest and most practical solution may be to place one of the cats in another home.

DOMINANCE AGGRESSION IN THE CAT

The social status between cats is intimately linked to their territorial—natures but in a very special way. Dominance between two or more cats is a function not only of the individuals but also the time of day and the place in which they interact. This is a type of social order called spatio-temporal dominance. When several cats are housed together or when the neighborhood population of strays and free-roaming pet cats rises, social pressures also increase. Under more crowded conditions, the use of a given territory is modified by the cat population to minimize aggressive encounters and maximize the exploitation of natural resources. Territorial conflicts will determine which cat gains preferential use of an area at a given time of day. The outcome of these conflicts also determines how the individuals will interact should they meet again. In many cases, the clear victor will be granted dominance and will be given the right of passage by the loser in future encounters. The loser, of course, has the option to patrol the same path but at another time of day and so avoid his or her rival.

A cat that establishes dominance over another cat along a given territorial boundary may be subordinate to other individuals somewhere else or at another time of day. For this reason, dominance and territorial aggression are intimately linked.

Over time, some cats patrol the same boundaries within an area but at different times of day while others follow different pathways of the territory at the same time of day. In some cases, individuals may patrol distinct areas but at the same time and meet along a common stretch of overlapping borders. The social dominance of cats is thus a more dynamic process compared to simpler linear hierarchies in other animals.

Spatio-temporal dominance is seen in housecats, too. When two or more cats share a home, they often will use different parts of your home at different times. One cat may sleep on top of the refrigerator while another moves about. Later, they alternate activity patterns so that the one plays in the den while the other sleeps on a favorite windowsill. Cats can develop lasting alliances and stay close by each other for much or part of the day. In other cases, they may merely tolerate each other. Should they meet in a doorway or hallway, for example, observant onlookers may notice which cat is given leeway. At these encounters, friendly greetings can be subtle and brief. When housemates are rivals, however, territorial conflicts with ultimate social significance may erupt.

AGGRESSION AND THE LITTER BOX

It should be noted that territorial anxiety can be expressed by undesirable behaviors other than aggressiveness. Cats mark their territories by leaving visible (scratch marks) or chemical (urine, stool or specialized scent glands on the skin) calling cards to relieve their own anxiety and signal others of their presence. In both indoor and outdoor pets, territorial tension frequently leads to increased marking behavior, including the deposit of waste in undesignated areas. This can also happen when territorial conflicts occur at or near the litter box, so that an individual additionally acquires an aversion to using the box. A cat may accidentally be startled, or it may be ambushed by a territorial rival taking advantage of an unguarded moment. Inappropriate elimination (voiding outside the box) is a non-specific response to anxiety in the cat and can be anticipated whenever and wherever territorial conflicts flare. Additional litter boxes should be provided at alternative locations to minimize territorial confrontations.

Fear aggression. Photo by Dr. Gary Landsberg.

REDIRECTED AGGRESSION BETWEEN PET CATS

If a cat is frightened or startled by something (such as an intruding cat outside the window), it may retaliate against the nearest available target. Frustration, redirected onto an uninvolved yet available individual, is fairly common. In a multi-cat household, the displaced target may unfortunately be another housecat. Whether the victimized housemate is witness to the triggering event or is unaware of its housemate's aggressive arousal, the attack may begin a cycle of aggressive interaction. Redirected aggression is, by definition, swift and intense. It can occur between housemates that, until then, peacefully coexisted. Sometimes, the dynamics between housemates quickly returns to normal but, more typically, it may take several hours, days or weeks until tempers subside. In some instances, the upheaval in the dynamics between cats will not resolve without professional intervention.

Following the initial eruption, a cycle of aggressive behavior is established. The target cat learns to anticipate an impending attack as soon as it sees the aggressor. Even if no attack was intended, the aggressor may learn to attack the other on sight, triggered by its defensive body postures and related signs of fear. Long after the initial episode, the actively aggressive cat may pursue its victim whenever it appears. Over time, the intensity of the chase diminishes and may even look like a kind of game. One of the common consequences of the social tensions is a renewal of territorial conflict. The dominant cat can actually appear to enjoy its reign of terror, but its victim's behavior is usually a clear indication that the interaction is not a playful one. Rarely, the roles evolve over time and may reverse so that the aggressor becomes the target cat.

In most cases, the owner will never know exactly what or whom set off the cat. Whenever possible, the identified trigger should be controlled or eliminated. This might require blocking access to a particular door or favorite window perch, for example, to prevent the housecat's view. Whenever possible, the cats in question should be immediately isolated from each other to avoid any further escalation of hostilities. Remember that redirected aggression can be redirected again toward you so proceed very cautiously. If the catfight is intense, it is safer to avoid injury to yourself and deal with any injuries sustained by your pets when the battle is over.

It is frequently necessary to keep both cats confined for about one week, each with its own food and water and litter pan. The use of anti-anxiety medication for either or both cats may be advised by your veterinarian. Because one or both cats are in a state of anxious arousal, psychoactive drugs can help to prevent a continuing vicious cycle and should continue until both cats have returned to normal interaction. When both appear calm and content in seclusion, you may let each one out by itself to readjust to your home. Over successive days, alternating periods of freedom can gradually become longer. If any sign of anxiety remains, extend periods of confinement by another day. When both cats seem comfortable, they may be briefly reintroduced at a scheduled mealtime. At first, they should be fed at opposite ends of the room. If there is any aggression, remove the food and separate them for at least 15 minutes or until the next scheduled feeding before trying again. Delay reintroduction if necessary and fall back to alternating confinement for as long as necessary. Premature reintroduction can prolong the process. In time and with your patience, things should return to normal.

AGGRESSION AND SEXUAL BEHAVIOR

Sexual maturity in female cats is signaled by the start of "estrus." The estral cycle is repeated at approximately three-week intervals. The fertile phase lasts about one week and repeats every two weeks during "mating season," between January and November. Under the influence of sexual hormones, the female cat, or queen, becomes increasingly agitated and vocal. She may sleep and eat less than usual though her restlessness heightens. Weight loss is common. She may try to escape outdoors. She may roll on her back, frequently licking the vulva, and solicit more attention from her owners. In the absence of a prospective mate, she may present her hindquarters with tail quivering to her owners. Owners may initially find all this amusing until their sleep is repeatedly disturbed by the insistence of a preoccupied estral queen.

Some aggressive postures are acceptable forms of cat play. Photo by D. Barron.

Estrus is a stressful and exhausting experience for the female. She may have insufficient time to recover before the next cycle begins. Not surprisingly, temperament changes, including lethargy and irritability, are common in cats between peak estral phases. Irritable aggression is also common in intact male cats, or tomcats.

Sexual drive in intact males naturally leads to escape attempts (if confined as housecats) and roaming. During mating season, territorial conflicts increase and are more likely to end in combat. Among the most intense and uninhibited types of aggression are the fights between intact toms when sexual tension is high.

Estral behavior in the queen is interrupted when mating triggers ovulation. Stimulation of the vaginal surface by tiny spiked ridges on the tom's penis triggers the brain to secrete hormones responsible for the release of eggs from the ovaries. Induced ovulation distinguishes the queen from ovulation in other species, such as human beings, who ovulate cyclically in the absence of sexual intercourse. Immediately after the male withdraws, the queen swiftly but briefly becomes intensely aggressive toward him. Soon after, the queen typically rolls and rubs against the ground and runs off. Post-coital aggression in the queen has been attributed to possible pain caused by the irritating surface of the penis during mating. It is also possible that the seemingly sudden burst of hostility is the more normal social behavior between unfamiliar cats. Aggressive outburst in the queen may reflect centrally inhibited aggression that lasts just long enough to allow the tomcat's approach and successful mating, only to rebound.

Aggressive patterns motivated by reproductive urges are best controlled by neutering. Surgical sterilization results in the immediate elimination of sexual hormones from circulation. If aggressiveness continues in pet cats following neutering, the type of aggression is either unrelated to sexual behavior or is governed by mechanisms other than only sexual ones. Fighting and roaming in male cats, for example, will be reduced following castration but are likely to continue as long as the opportunity is presented. This is partly because territorial behavior, made more intense by underlying sexual drives, has an additional function in providing resources such as food, shelter and water, which are survival basics.

AGGRESSION AND MEALTIME

Anxiety at mealtime may be expressed as over-eating or eating too quickly and is seen in pets that have never been deprived of food. Anxiety in the form of aggressive behavior can be directed toward the pet's owner or toward another pet. Aggressive behavior associated with feeding is not a sign that a pet ever experienced starvation. Irritable aggression because of hunger is one explanation. Competition over food can instigate aggressiveness in multi-pet households. Guarding, sometimes called "protective aggression," is another type of aggression seen in cats and other animals to defend an object of value such as food.

Competition between pets over

food is usually easy to defuse by feeding them at separate locations. If it is not enough to feed them at different places in the same room, it may be necessary to isolate one or both or all while they eat. If you are currently feeding your pets only once a day, divide the daily portion into at least two daily meals so that the anxiety is not so focused.

AGGRESSIVE PLAY IN CATS

There are few sights more amusing than a cat at play. Play, however, is serious business to cats at any age. Particularly important for kittens and young adults, playful behavior allows the individual to learn about the world and its place in it. Play provides the individual with important life skills with relatively few serious consequences. It can learn about its territory, for example, and its physical strength and social status relative to other cats. Playful games of chase practice vital skills necessary to escape real danger. Other playful behavior gives the cat practice for successful hunting maneuvers. For juvenile and adult cats alike, play functions to maintain flexibility, muscle tone, and general fitness.

Play is also one of the best ways to teach a pet acceptable behavior. This is particularly true for kittens. Forms of play that encourage or even tolerate a cat's aggressiveness during play may convey the unintended message that aggressive behavior is acceptable. Cats and kittens alike should be discouraged from nipping, scratching, pouncing on or hanging from any article of clothing or body part belonging to any person. A common but inappropriate game that some naive pet owners play is to encourage their cats to "attack" wiggling fingers or feet (beneath blankets, for example). Cats are naturally attracted to moving targets. A playful cat's attention should be directed toward a target that is apart from its human playmate. Provide a wide variety of toys that can be thrown by you and chased by your cat. Other toys may be dangled from string or wire or at the end of a stick. Your cat can learn to express its basic instincts without harming you.

Invite your cat to play with you instead of waiting for it to seek your attention. Left to its own devices, you may not like what it learns to get you to notice it. Some cats will learn to get your attention by nipping at your heels, for example. Engage your cat on a daily basis in appropriate forms of play so that it need not resort to undesirable yet effective ways to attract your attention.

Cats learn a lot about each other when they play together. During playfights, each learns about the other's stamina, strength and assertiveness. This type of information can be useful when territorial disputes occur. Sometimes, territorial disputes are expressed in what appears to begin as play or are mistaken by observers as innocent play. Play times may end abruptly if one cat gets too rough, but this may also be a thinly veiled territorial conflict or challenge for dominance.

Playfulness varies between cat breeds and individuals within each breed. Kittens are famous for their playful exploits, but the intensity and proportion of playtime decreases with age. While some adult cats are not very playful, healthy cats often continue to play for most of their lives. In a usually playful cat, any decrease in playful activity may be a sign of illness. When aggression is seen in a normally even-tempered cat, the possibility of an underlying medical problem should be investigated. In fact, unusual aggression associated with any activity should be monitored and discussed with a veterinarian.

PREDATORY AGGRESSION IN THE CAT

With the exception of lions, most cat species hunt alone and at night. The cat's reliance on vision to detect and capture its prey is reflected in the size of its eyes relative to its body size, the greatest proportion of any predatory mammal. A specialized layer of tissue at the back of its eye serves to amplify even dim lighting. At closer range, other sensory information from smell, vibration and hearing becomes increasingly important. The healthy cat is swift, agile and focused. Equipped with the perfectly appropriate weaponry of its teeth and claws, the cat is a supremely qualified hunter.

Not every cat is born with strong hunting instincts. In fact, predatory behavior is only partly instinctive; it must also be learned. Kittens perfect inborn hunting skills by observing their mother in action and practice elements of predatory patterns during play. Individual preference for a particular type of prey is also influenced by local availability. In some

Sometimes, competition over food can cause problems in multi-pet households. If this happens, you can set up different feeding "stations" in the room. If that doesn't work, you can isolate each pet at mealtime. Photo by Robert Pearcy.

housecats, the instinct to hunt may be so strong that they are immediately successful without prior experience.

For many cat owners, their cat's hunting prowess is anything but desirable. The sight of a still-living or half-eaten mouse dangling from a cat's mouth can be horrifying. While some cat owners interpret the presentation of prey as a gift of gratitude, it is more likely an inborn or learned behavioral sequence. This might also reflect a basic instinct to return to the "den" or other secured area to devour a meal.

A bell attached to a break-away collar can sometimes give adequate warning to otherwise unsuspecting victims, unless the cat learns to move without ringing the bell. Two or more bells on a collar may improve the plan. Surgical removal of a cat's front claws, commonly referred to as "declawing," is not recommended to control unwanted predatory aggression. Most declawed cats adapt quickly and continue to hunt successfully. Declawed cats are at a defensive disadvantage and should be strictly confined as housepets.

The only realistic way to stop cats from hunting is to prevent the opportunity for hunting. Regardless of how well it is fed at home, cats that are permitted to roam outside will fulfill their predatory drive. A one-way cat door that allows your cat to exit freely but restricts its return might allow you to screen for unwelcome visitors. Allowing your pet the freedom to roam is a high-risk decision. In addition to a multitude of health hazards, predatory cats are in jeopardy of being seriously injured by their intended victim and exposed to contagious diseases.

SUMMARY

Aggressive behavior is among the most important survival mechanisms for many animals, and cats are no exception. There are many forms of aggression. To interpret the reason for any aggressive display, the details of its context and a history of the individuals(s) involved are necessary. This chapter is a review of some of the more common aggressive types seen in pet cats, and some simple remedies for their owners.

ADDITIONAL READING

Borchelt, P.L.; Voith, V.L.: *Aggressive Behavior in Cats.* Compendium on Continuing Education for the Practicing Veterinarian, 1987.

Borchelt, P.L.; Voith, V.L.: *Diagnosis and Treatment of Aggression Problems in Cats.* Veterinary Clinics of North America, Small Animal Practice, 1982.

Neville, P.: *Do Cats Need Shrinks? Cat Behavior Explained.* Contemporary Books, Chicago, 1990.

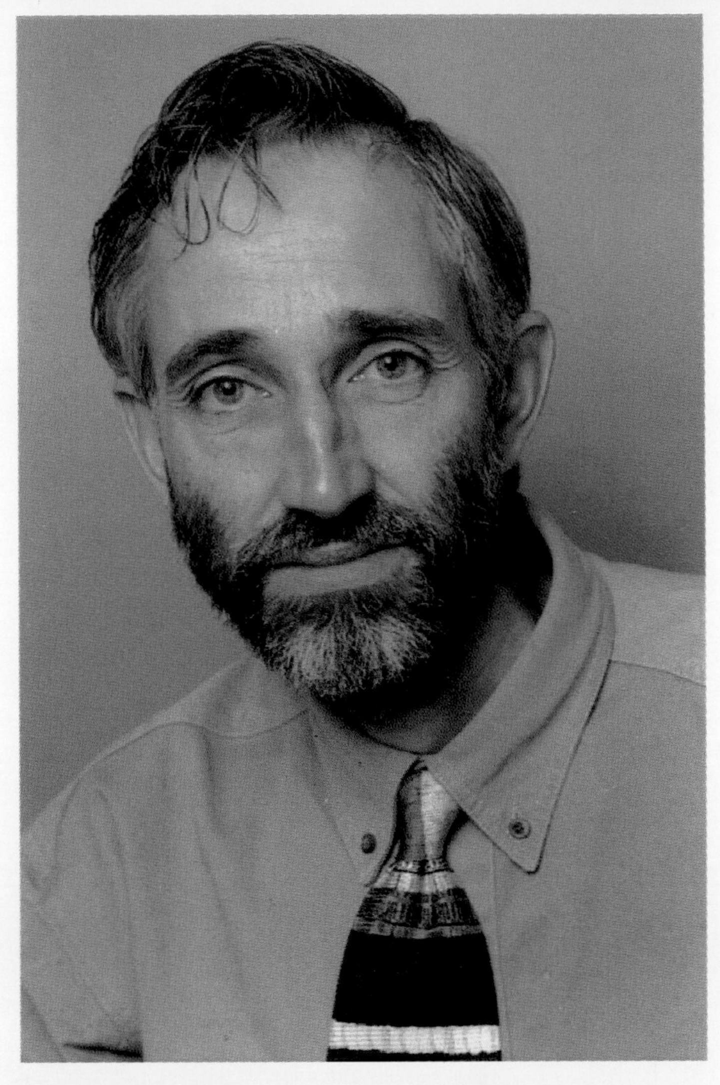

Dr. U.A. Luescher graduated with his doctor of veterinary medicine degree from Zurich in 1979. In 1984 he completed his PhD from the University of Guelph and got his professional certification from the Animal Behavior Society in 1992. He is currently an assistant professor of ethology at the Ontario Veterinary College in Guelph, Ontario, Canada. Dr. Luescher is a Diplomate of the American College of Veterinary Behaviorists.

Compulsive and Conflict-related Problems

By Dr. U.A. Luescher
Department of Population Medicine
Ontario Veterinary College
University of Guelph
Guelph, Ontario, N1G 2W1
Canada

INTRODUCTION

A cat curled up in a sunny spot is the personification of contentment and relaxation. However, many cats are quite susceptible to stress due to even minor changes in their environment and have great difficulties coping with these changes. Many cats will spray urine or eliminate outside the litter box in response to stress, but some may start to perform seemingly very unnatural and maladaptive behaviors, sometimes even resulting in self-injury. Of all the cats that are treated for behavior problems at the Ontario Veterinary College, about 7% are affected by a stress-related condition known as compulsive disorder.

There can be various sources of stress in a cat's life. They usually involve relationships with other cats or with people, a change in schedule, or any change in the environment of the cat. How much a cat is affected by these factors and how the cat responds to them depend to a large degree on the cat's genetic makeup.

FORMS OF COMPULSIVE BEHAVIOR IN CATS

The most common compulsive behavior which cats exhibit relates to grooming. Some cats may lick themselves excessively until they have bald spots and possibly raw skin. Others will chew their hair off. The most commonly affected areas of the body are on the topline, the belly and between the thighs. Sometimes, it may be difficult to sort out whether the changes are caused by a behavior problem or by a skin disease.

Some cats, particularly Siamese, Burmese, and their crosses, may suck on woolen objects such as sweaters or blankets. Others may prefer other materials such as rubber or plastic. In extreme cases, cats may ingest large parts of these objects.

Some cats can relax, but many are susceptible to stress due to even minor changes in their environment. Photo by Robert Pearcy.

Stress can also be expressed in sudden agitation and abnormal movements such as head flicking, very rapid licking, dashing away, pouncing on imaginary prey, or maintaining a frozen body position. Usually, when agitated, these cats appear to have a glazed look due to their pupils being dilated, and a rippling of the skin over the back can be observed.

Some cats seem to be hallucinating and to be reacting to imaginary objects. A cat may, for instance, repeatedly look up and then duck, as if an object was flying by just above its head.

The most disturbing compulsive behaviors are those which lead to self-inflicted injury. Cats may direct severe aggression towards parts of their bodies, usually their tail, and bite themselves very hard to the point of drawing blood.

CAUSES OF COMPULSIVE BEHAVIOR IN CATS

Compulsive behavior may have many causes, but it is assumed that it most commonly is induced by chronic stress or conflict. Conflict in cats might result from a very restrictive environment, e.g., crate confinement, in which the cat cannot express such species-typical behavior as prey catching or scratching. In some cats, social isolation may result in a conflict as well, whereas others will prefer to be alone most of the time and are bothered by too much attention.

Fear without the possibility to escape induces conflict and is very stressful. Cats may be fearful of noises, e.g., originating from appliances, of people, or of other pets in the household. If an animal is fearful of people, but at the same time would like to socialize, it is in a state of motivational conflict. Sudden aggression when cats are petted for prolonged times could be resulting from a conflict between wanting to be on the owner's lap, and not wanting to be petted.

Cats are very sensitive to change. Moving furniture, redecorating, changes in the owner's schedule and, maybe most importantly, changes in the social group in which the cat lives, all are very stressful to many cats.

Tension within the cat's social group obviously contributes to stress. If a person doesn't like the cat, if a cat is fearful of people because it wasn't sufficiently socialized as a kitten, if a cat which would prefer to be left alone is handled excessively, e.g., by children, or if the owner uses a spray bottle or other means to punish the cat, these could all result in the cat displaying compulsive behavior. The presence of other cats in a multi-cat household or a strange cat visiting the yard or sitting on the window sill, may also lead to social tensions and stress.

If and how a cat expresses stress depend on its genetic makeup. Some cats are genetically very easy going and friendly, and feel comfortable in any situation. Others are more easily disturbed and unless very well socialized as kittens, will be loners and do not take well to social contact. Cats may also express stress, even chronic stress, in various ways, and, again, this depends largely on inheritance. Most cats would spray urine to mark their territory with their own smell, so they will feel more secure and more comfortable. Others, however, would develop one or another form of compulsive behavior. A genetic influence on how cats express stress is clearly demonstrated by the fact that woolsucking is much more common in Siamese and Burmese than in any other breed of cat.

Every cat will occasionally show signs of conflict. If the stressful situation is brief and does not occur very often, this should be of no concern to the owners. If, however, the conflict is long-lasting and the cat cannot resolve it or escape from it, or if the stressful situation repeats itself frequently, this may have an almost permanently damaging effect on the cat. In these cases, there are biochemical changes taking place in the brain. These changes will not readily reverse, even if the stress eventually is eliminated. These changes in brain chemistry are what make the cat engage so compulsively in the very strange behaviors described above.

PREVENTION AND TREATMENT OF COMPULSIVE BEHAVIOR

The first few weeks of life are of paramount importance to the development of puppies and kittens. Experiences during that time have long-lasting effects on the disposition of the adult dog or cat. Kittens and puppies experience the same things, although this phenomenon has been less well researched for cats than for dogs. Kittens which are handled regularly during their first days of life will not only develop faster and be healthier but will also be less susceptible to stress later in life. Between about two and six weeks of age, kittens go through a so-called socialization period. The more contact they have with people during that time, the friendlier they will be. (How much socialization to people an individual needs is an inherited trait, passed on mostly by the father.) The more a kitten is exposed to different new situations, the more different experiences it acquires during that time and the easier it will adjust to the various situations it will encounter in its later life.

Once a compulsive behavior is established, the first step in treatment is to identify and to try to eliminate the source of stress. If it cannot be eliminated, maybe it will be

Overgrooming of the haircoat, so-called psychogenic alopecia, is one manifestation of compulsive disorder in cats. Photo courtesy of Dr. Lowell Ackerman.

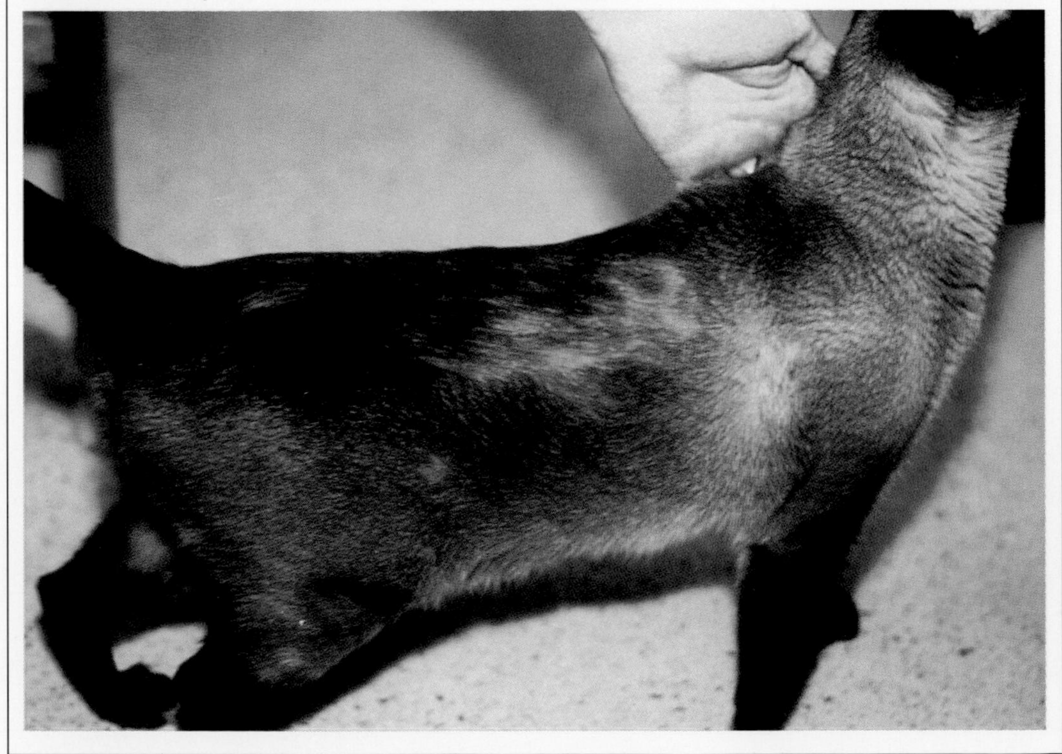

Sometimes it's difficult to determine the cause of a cat's stress. Photo courtesy of Dr. Gary Landsberg.

possible to gradually get the cat used to it, i.e., to desensitize the cat to whatever causes the stress. Furthermore, we always recommend that the cat be ignored as much as possible, with the exception of some quality time given on a regular basis, e.g., one hour every evening from six to seven o'clock. Ignoring the cat has two functions. First, the owners will not give the cat any cues which the cat associates with a conflict situation, and second, the owners will not inadvertently reinforce the compulsive behavior by paying attention to the cat when it displays the behavior. We also strongly advise that no punishment be used with the cat ever again. Punishment such as spraying the cat with water, yelling at the cat, or even hitting it, is usually a source of severe stress, because the timing of the punishment is usually imprecise, the intensity may be inappropriate, and punishment can usually not be applied consistently. It also makes the cat fearful of the owner. Any interaction with the cat, such as feeding and of course, play time, should be on as consistent a schedule as possible.

If a compulsive behavior has been going on for some time, the levels of certain neurotransmitters, or chemicals in the brain, may be altered. At this point, it is usually necessary to correct this chemical imbalance with a drug. However, the drug should be considered only as part of the therapy, not as a cure on its own.

In very persistent cases, it may be necessary to supervise and consistently distract the cat whenever it exhibits the compulsive behavior. Care has to be taken that the distraction occurs just as the cat is about to perform the behavior, and that it is not perceived as pleasant by the cat, because otherwise it may be reinforcing the behavior.

SUMMARY

Compulsive behavior can be a very severe behavioral disorder and is very distressful to the cat and its owner. Many cases can, however, be treated successfully, particularly if they haven't been going on for too long.

ADDITIONAL READING

Overall, Karen L. "Recognition, Diagnosis, and Management of Obsessive Compulsive Disorders. Part 1: A Rational Approach." *Canine Practice*, 1992.

Overall, Karen L.: "Recognition, Diagnosis, and Management of Obsessive Compulsive Disorders. Part 2: A Rational Approach." *Canine Practice*, 1992.

Overall, Karen L.: "Recognition, Diagnosis, and Management of Obsessive Compulsive Disorders. Part 3: A Rational Approach." *Canine Practice*, 1992.

Facing page: How much a cat is affected by the stressors in its environment depends to a large degree on the cat's genetic makeup. Photo by Robert Pearcy.

Caroline B. Schaffer is Director of External Affairs at Tuskegee University's School of Veterinary Medicine in Tuskegee, Alabama. She also serves as advisor to Tuskegee University's Human-Animal Bond/ Animal Behavior Club and the Student Chapter of the American Association of Feline Practitioners. She supervises the University's volunteer program that prepares veterinary medical students and their pets to work in pet-facilitated therapy programs. Prior to moving to Tuskegee University, Dr. Schaffer worked as a research associate in small animal surgery at Iowa State University's College of Veterinary Medicine and practiced small animal medicine and surgery in group practices in a suburb of Chicago, Illinois, and in Columbus, Ohio. She was honored by the Association of Teachers of Veterinary Public Health and Preventive Medicine with their Michael J. McCulloch, M.D. Memorial Award in 1994 "in recognition of her outstanding contributions in research, teaching, and service related to the human-animal bond." She earned her Doctor of Veterinary Medicine degree from The Ohio State University College of Veterinary Medicine in 1971.

David D. Schaffer is currently a resident in psychiatry at the University of Alabama at Birmingham. He received his DVM from Tuskegee University's School of Veterinary Medicine in Tuskegee, Alabama, in 1971. After practicing small animal medicine and surgery in Columbus, Ohio, he returned to academia. He earned a PhD in pharmacology from Loyola University of Chicago in 1980 and completed a post-doctoral fellowship in clinical pharmacology at the University of Chicago in 1982. Thereafter, he did research in veterinary pharmacology and taught veterinary pharmacology and physiology at both Iowa State University's College of Veterinary Medicine and Tuskegee University's School of Veterinary Medicine. He helped to establish Tuskegee University's Human-Animal Bond/ Animal Behavior Club in 1989-90. He earned a Doctor of Osteopathy degree at the University of Osteopathic Medicine and Health Sciences in Des Moines, Iowa in 1993. His diverse professional training merges his interest in and respect for the human-animal interdependent relationship.

Sometimes the Best Therapist is a Cat

By Caroline Brunsman Schaffer, DVM
David D. Schaffer, DVM, PhD, DO
Tuskegee University
Tuskegee, AL 36088

A thin man with a twisted arm softly mouths the words, "Pretty kitty," and the nursing home staff stare in amazement. They tell the volunteer visiting with her tabby cat that the gentleman hasn't talked for months. A veteran without legs propels his wheelchair from room-to-room and up and down the hallway of a Veterans Affairs Medical Center just to be near the chunky orange and white cat visiting the nursing home residents that day. At one point, he tells the volunteer about his experiences as a paratrooper in the war. By contrast, when therapy dogs visit him, he stays in the Day Room, politely gives the dogs a quick pat, and speaks only long enough to ask for a cigarette.

An autistic child stops her chronic rocking and focuses her attention on the Tonkinese cat sitting on her lap. She remains motionless as she talks to the cat. The volunteer smiles, knowing that for several minutes, the young girl is in touch with the world about her.

INTRODUCTION

For every pet-facilitated therapy program there are stories of the beneficial effects that cats have on people with physical, emotional, or behavioral problems: stories about people who haven't talked for months or even years until a cat came to visit, about nursing home residents who are happier and more sociable when a cat is rubbing against them, about hospitalized patients who need less pain-relieving medications when a pet is present, about cats who help with physical therapy sessions by encouraging patients to stroke or brush them, and about care givers who are more attentive to their patient's needs when a cat or other pet comes to visit.

Promoters of animal-assisted activities talk about the enhanced human interactions that occur when owners see how readily their cats accept people regardless of their physical appearance or mental state. Some pet owners, previously uncomfortable in institutions, have become very effective and dedicated volunteers simply because of their eagerness to share their pet's love with a stranger.

GROWING ACCEPTANCE OF CAT THERAPY

Recognition of the diverse benefits of people-pet interactions has contributed to growing acceptance of pet-facilitated therapy and other animal-assisted activities by physicians and nurses, psychiatrists and psychologists, physical and occupational therapists, social workers and educators, and institutional administrators and trustees throughout the world. After years of being barred from entry into hospitals, nursing homes, and schools because of fears of health risks to people from injuries or zoonotic diseases, i.e., infections shared by people and pets, cats and other companion animals are now being welcomed with open arms. Having a resident cat (one who lives at the facility) or a visitation cat (one who comes to the facility on scheduled visits with its owner) has become so popular that some institutions even advertise that they offer pet-facilitated therapy when they are recruiting health care providers.

Fears of major health risks have not materialized. In a study of 233 skilled and intermediate care facilities in Illinois, for example, only 24 facilities reported any safety problems over a one-year period. In another 12-month study, 284 Minnesota nursing homes found no incidence of infections spread by pets and only 19 mechanical injuries–only two of which included broken bones–due to pets as compared with many injuries due to falling in the bath or out of bed.

Although it is true that animals appear to be working miracles through pet-facilitated therapy, not all cats are miracle workers. Contrary to popular belief, animals don't always give unconditional love. In addition, not all people

Persian kitten displaying all the charms of its breed. Photo by Robert Pearcy.

Design of PUPS Test

Tuskegee University
School of Veterinary Medicine
Tuskegee, Alabama

Waiting Room: Interaction of pet with other animals in a room full of strangers
1. Pet and handler stay in a large room with a window where other pets and their handlers are mingling as they wait to be tested
2. The activities will be unstructured, but evaluator will observe pet's reaction when approached by strangers and by animals
3. Evaluator will observe pet's reaction when it approaches strangers and other pets
4. Evaluator will observe pet's reaction to activity outside the window

Examination Room: Approached by stranger/Manipulations by strangers
1. Pet and handler enter examination room
2. Two evaluators (one male and one female) enter examination room
3. Handler puts pet on table
4. Evaluators pull on pet's feet, ears, tail; lift its lips; stroke its hair roughly; look directly into its eyes; move it from side-to-side; talk to it in rough voice, normal voice, and squeaky voice

People Hall: Willingness to approach and be with strangers
1. Pets and handlers walk down hallway containing several strangers who are standing, walking, and sitting. If pet does not walk on a leash (ex: rabbit, Guinea pig, cat), then the pet will be carried through the hallway of people by its handler
2. Handler and pet stop to greet first stranger who shakes hands with handler, stares directly at pet, pats the animal on its head and shoulders, grabs handler's arm, and grabs pet's leg.
3. Handler and pet move down hallway and are encircled by strangers who pet the animal, move erratically, and talk excitedly
4. Handler puts lap-sized pet on lap of seated stranger and hands leash of larger pet to stranger. Handler tells pet to stay and then walks away from pet but within view of the pet for 30 seconds. After 30 seconds, stranger either carries pet or walks with pet through the hallway to the handler.

Distractions: Response to people and animals outside pet's reach
1. Pet and handler move to glass door.
2. Strangers walk past the door within view of the pet
3. Some strangers will have pets with them
4. If pet is on leash, handler may ask the pet to sit or stand. If pet does not walk on a leash (ex: rabbit, Guinea pig, cat), then the pet may be held by the handler

Obstacle Course: Tolerance of various movable and stationary objects
1. Pet and handler move through an obstacle course of stationary and movable objects while evaluator stands off to the side. If pet does not walk on a leash (ex: rabbit, Guinea pig, cat), then the pet will be carried past the obastacles by its handler
2. A cart on wheels (preferably one with squeaky wheels) will be pushed past the pet
3. The pet will move past a tray of food, a bed, and a stranger with a cane
4. An umbrella will be opened and closed rapidly near the pet
5. Depending on the pet's size, the pet will be placed on the bed
6. Stranger will lead or carry pet through the obstacle course after the handler has taken the pet through the course

Elevator/Small Enclosure: Claustrophobic tendencies
1. Handler leads pet into small, dark room already occupied by the evaluator. If pet does not walk on a leash (ex: rabbit, Guinea pig, cat), then the pet will be carried into the small room
2. Evaluator closes the door and turns on the room's light
3. Handler, pet, and evaluator stay inside the small room for 60 seconds

Noise Hall: Tolerance of noxious sounds
1. Pet and handler stand in hallway while various noises are generated around them. Animal may sit, stand, or be held
2. The stranger will make noises by clanging dishes, ringing a bell, speaking in a loud voice, dropping books and trays, shuffling and stomping his/her feet
3. A buzzer or vibrator will be held close to animal

Simulated Day Room: Interaction of pet with other animals and a room full of strangers
1. Pet and handler enter a room where strangers are seated and other pets and their handlers are mingling
2. Pets will be encouraged to move past one another as they stop to greet each stranger who is seated in the room

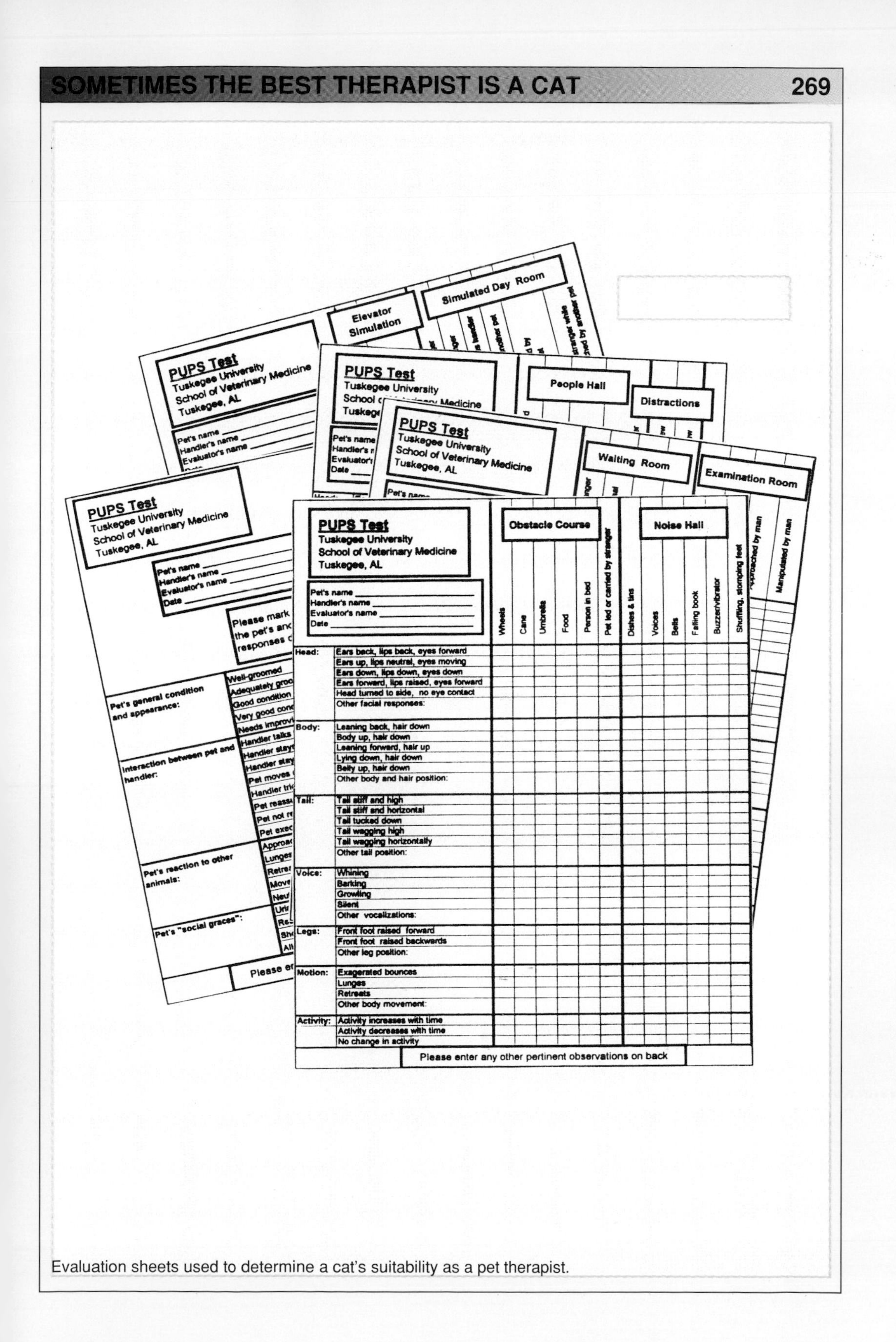

PUPS Test
Tuskegee University
School of Veterinary Medicine
Tuskegee, AL

Pet's name ______
Handler's name ______
Evaluator's name ______
Date ______

		Obstacle Course						Noise Hall					
		Wheels	Cane	Umbrella	Food	Person in bed	Pet led or carried by stranger	Dishes & tins	Voices	Bells	Falling book	Buzzer/vibrator	Shuffling, stomping feet
Head:	Ears back, lips back, eyes forward												
	Ears up, lips neutral, eyes moving												
	Ears down, lips down, eyes down												
	Ears forward, lips raised, eyes forward												
	Head turned to side, no eye contact												
	Other facial responses:												
Body:	Leaning back, hair down												
	Body up, hair down												
	Leaning forward, hair up												
	Lying down, hair down												
	Belly up, hair down												
	Other body and hair position:												
Tail:	Tail stiff and high												
	Tail stiff and horizontal												
	Tail tucked down												
	Tail wagging high												
	Tail wagging horizontally												
	Other tail position:												
Voice:	Whining												
	Barking												
	Growling												
	Silent												
	Other vocalizations:												
Legs:	Front foot raised forward												
	Front foot raised backwards												
	Other leg position:												
Motion:	Exagerated bounces												
	Lunges												
	Retreats												
	Other body movement:												
Activity:	Activity increases with time												
	Activity decreases with time												
	No change in activity												

Please enter any other pertinent observations on back

Evaluation sheets used to determine a cat's suitability as a pet therapist.

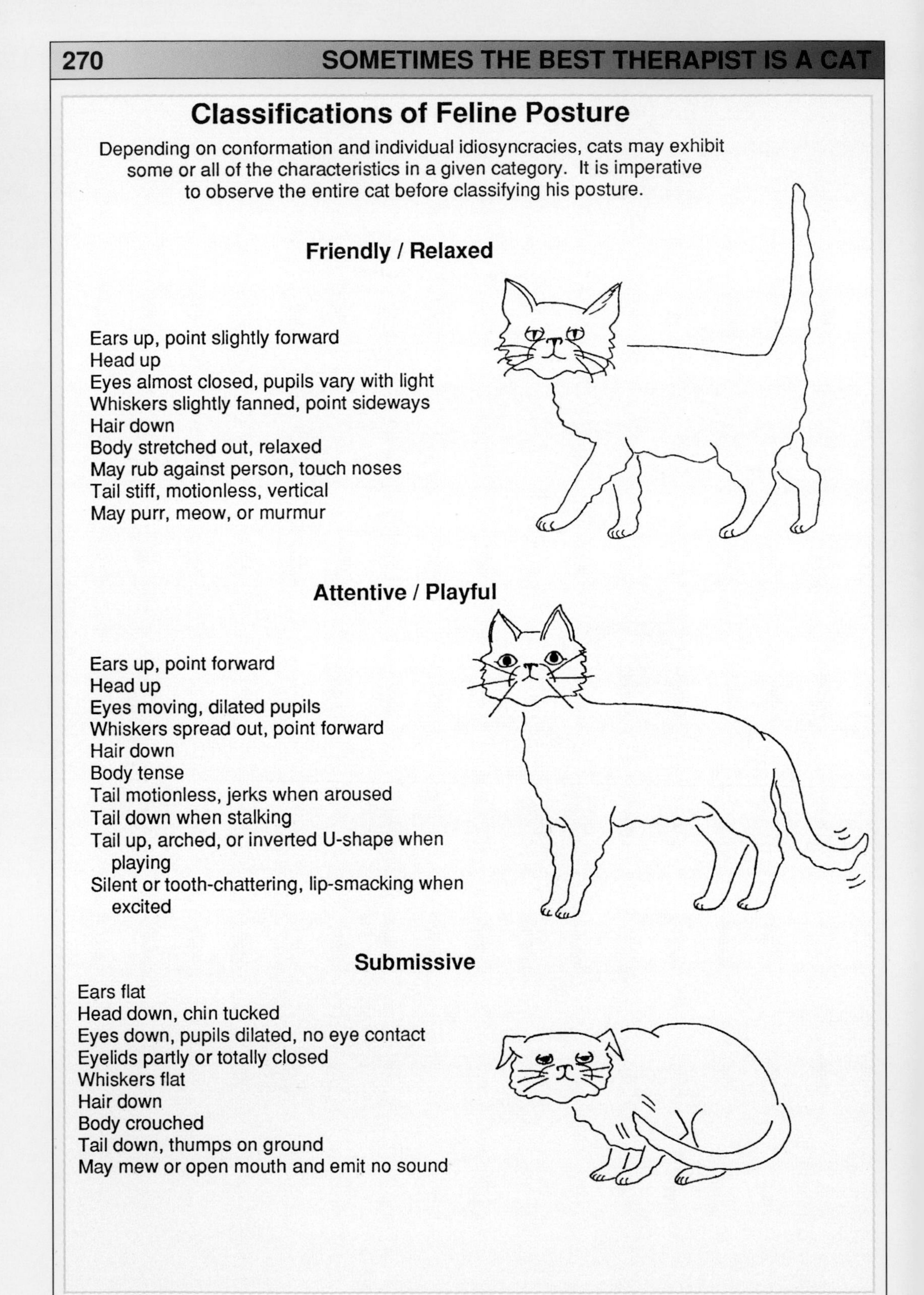

Classifications of Feline Posture

Depending on conformation and individual idiosyncracies, cats may exhibit some or all of the characteristics in a given category. It is imperative to observe the entire cat before classifying his posture.

Friendly / Relaxed

Ears up, point slightly forward
Head up
Eyes almost closed, pupils vary with light
Whiskers slightly fanned, point sideways
Hair down
Body stretched out, relaxed
May rub against person, touch noses
Tail stiff, motionless, vertical
May purr, meow, or murmur

Attentive / Playful

Ears up, point forward
Head up
Eyes moving, dilated pupils
Whiskers spread out, point forward
Hair down
Body tense
Tail motionless, jerks when aroused
Tail down when stalking
Tail up, arched, or inverted U-shape when playing
Silent or tooth-chattering, lip-smacking when excited

Submissive

Ears flat
Head down, chin tucked
Eyes down, pupils dilated, no eye contact
Eyelids partly or totally closed
Whiskers flat
Hair down
Body crouched
Tail down, thumps on ground
May mew or open mouth and emit no sound

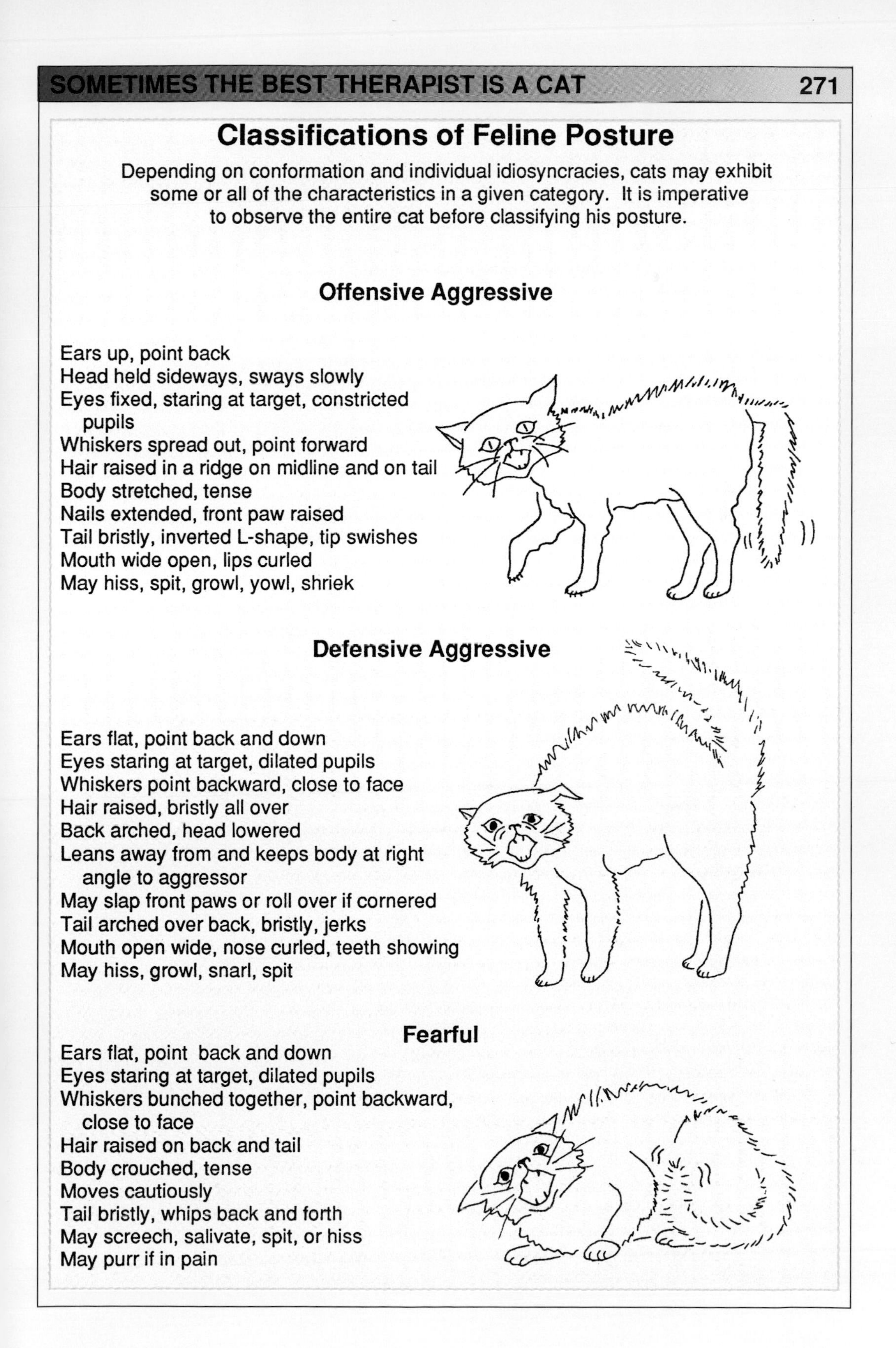

Classifications of Feline Posture

Depending on conformation and individual idiosyncracies, cats may exhibit some or all of the characteristics in a given category. It is imperative to observe the entire cat before classifying his posture.

Offensive Aggressive

Ears up, point back
Head held sideways, sways slowly
Eyes fixed, staring at target, constricted pupils
Whiskers spread out, point forward
Hair raised in a ridge on midline and on tail
Body stretched, tense
Nails extended, front paw raised
Tail bristly, inverted L-shape, tip swishes
Mouth wide open, lips curled
May hiss, spit, growl, yowl, shriek

Defensive Aggressive

Ears flat, point back and down
Eyes staring at target, dilated pupils
Whiskers point backward, close to face
Hair raised, bristly all over
Back arched, head lowered
Leans away from and keeps body at right angle to aggressor
May slap front paws or roll over if cornered
Tail arched over back, bristly, jerks
Mouth open wide, nose curled, teeth showing
May hiss, growl, snarl, spit

Fearful

Ears flat, point back and down
Eyes staring at target, dilated pupils
Whiskers bunched together, point backward, close to face
Hair raised on back and tail
Body crouched, tense
Moves cautiously
Tail bristly, whips back and forth
May screech, salivate, spit, or hiss
May purr if in pain

enjoy the companionship of animals. Some may even mistreat the animals who come to visit them.

JOB INTERVIEWS FOR CATS

So, which cats can help the sick, the physically or mentally challenged, the lonely, the dying, and the elderly? Just as a person must go through a screening process when applying for a job, so, too, should every potential feline therapist be selected with care. All cats should pass a behavior test that simulates the facility they will visit.

If cats are not selected for their good health and suitable behavior, the prejudice against pets in nursing homes and hospitals could return. The doors could, once again, be slammed shut in the faces of feline therapists. Tragically, many people would be denied the potential benefits that come from interacting with a cat because of the carelessness of those who did not participate in a comprehensive screening program.

The only way to know with certainty how a particular cat will respond to patients/residents at a given institution is to take him into that institution. However, the potential injuries to people and the potential stress on the cat that could occur if he reacted poorly in the novel environment make it safer to first test through simulations. This testing requires commitment, creativity, and time if it is to accurately assess the cat. The test must simulate as nearly as possible the facility where the cat will be working. It should include those sights, sounds, odors, and activities that the cat might encounter on any visit. If, for example, the cat is to visit people in wheelchairs, then he should be tested around wheels. If he is to visit children, then his interactions with children should be observed. If he is to work with other animals, then his interactions with each species should be observed.

People throughout the United States have designed behavior tests to help select animals for special tasks. Some institutions require proof that a pet has passed a specific test; others have no requirements. For example, some permit only registered "Pet Partners" i.e., pets who have passed the behavior and health screening test and handlers who have passed the written examination given by the Delta Society, to interact with patients and residents.

Regardless of the institution's requirements, cat owners would be wise to have their cats tested before taking them to an unfamiliar setting. Just as not all people enjoy visiting with animals, not all cats want to sit on a stranger's lap or go for a ride on a wheelchair. Not all will lie patiently for a tummy rub or allow their ears and legs to be pulled. Some cats are too aggressive. Others too fearful. Some may even be too playful. A good behavior test helps an owner determine the suitability of his cat for a

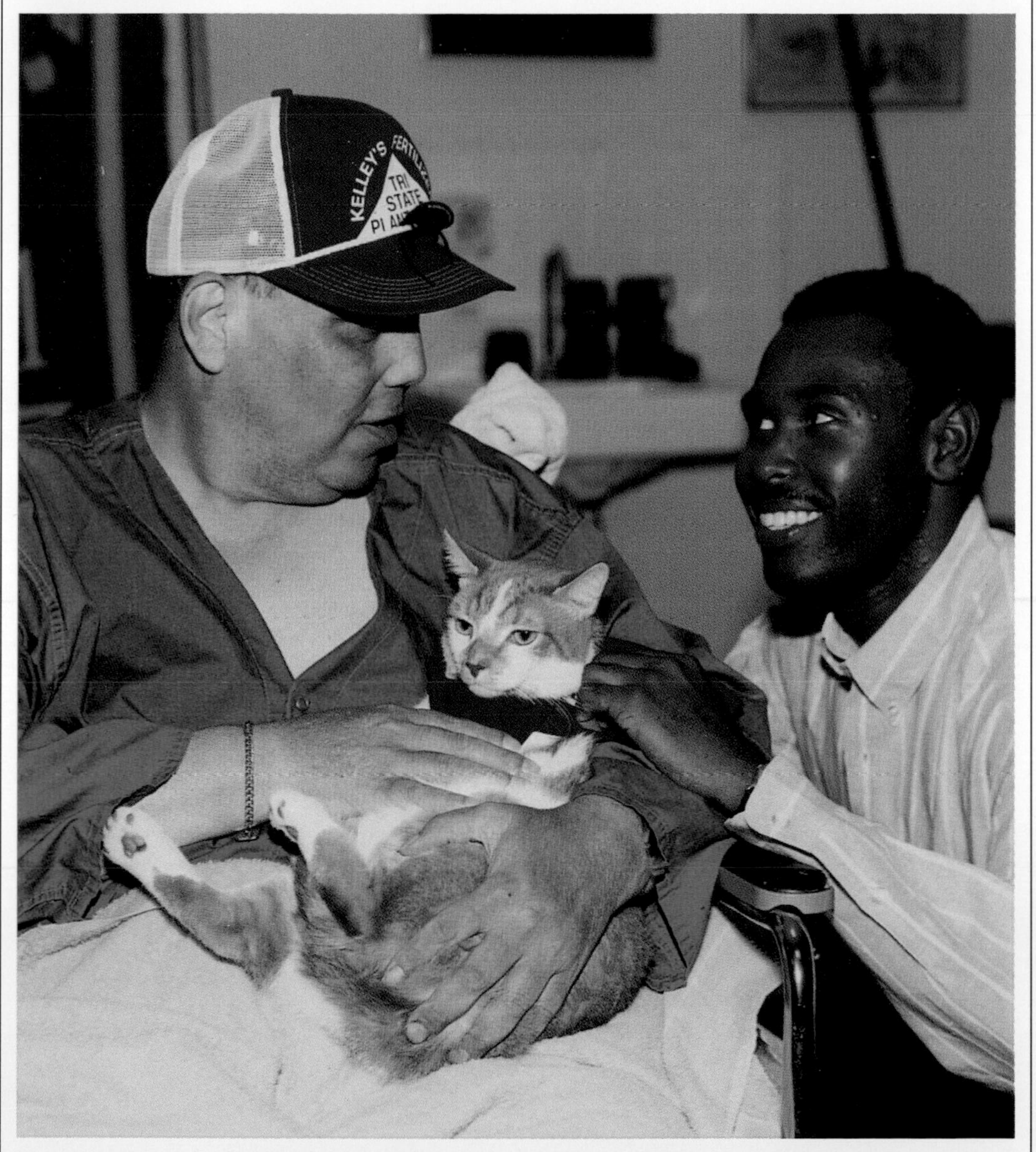

Some hospitals and other health-care facilities have implemented pet-therapy programs that have been of significant benefit to their patients. Photo courtesy of Dr. Caroline Schaffer and Dr. David Schaffer.

particular therapy program. It also gives him a better understanding of his cat's tolerance level. Ultimately, it is the owner's responsibility to be sure that his cat will behave appropriately and not be unduly stressed by the therapy program.

Most screening tests were designed for dogs, but they may be modified for cats. A few of the tests in use in the United States include the Tuskegee PUPS Test (acronym for Pets Uplifting People's Spirit) of the Tuskegee University School of Veterinary Medicine, Tuskegee,

Alabama; the Pet Visitation Screening Test of the San Antonio Delta Society, San Antonio, Texas; the Canine Good Citizen Test of the American Kennel Club, Raleigh, North Carolina; the Pet Partner Test of the Delta Society, Renton, Washington; the PAWS Temperament Test (acronym for Pets Are Working Saints) of the Frazier Memorial United Methodist Church, Montgomery, Alabama; the Temperament Test of the American Temperament Test Society, Inc.; and the Puppy Behavior Test of William E. Campbell as presented in *Behavior Problems in Dogs*, 1975.

A MODEL BEHAVIOR TEST

The Tuskegee PUPS Test is described in this chapter as an example of how a cat might be selected. This test was designed so that:

1. People with minimal knowledge of companion animal behavior could easily administer the test and evaluate the responses,
2. Activities at each station would not overwhelm the pet,
3. The test would be safe for the pet, handler, and evaluators,
4. The various situations that could be encountered during a nursing home/hospital visitation would be simulated,
6. Evaluations would be objective–not subjective,
7. The pet's interactions with other animals and with strangers of both sexes would be observed at more than one station, and
8. The behavior exhibited in the test would be a reliable predictor of behavior that would be exhibited in the arena in which the pet would be working.

To meet these eight criteria, the PUPS Test was designed in a grid format that asks the evaluator to describe the pet's body language. The veterinarians and veterinary medical students who created the test deliberately avoided subjective terms such as excited, shy, and happy. Instead, non-judgmental terms such as ears forward, tail stiff and high, and lips raised were used to describe the pet's posture.

The grid format was designed to minimize the amount of writing required by the evaluator. Key aspects of the eyes, ears, mouth, tail, etc. were selected as the elements that would be most expressive and easiest to observe when the test animal encountered another animal or a person. These elements were selected based on reports by animal behaviorists on the body language that animals use to communicate nonverbally with others of the same species. As shown by one of six postures (attentive/playful, friendly/relaxed, submissive, defensive aggressive, offensive aggressive, and fearful), the entire body must be observed before interpreting a cat's reaction at each station. Observing only one feature such as ear carriage or tail motion could be misleading.

Eight stations (Waiting Room, Examination Room, People Hall, Distractions, Obstacle Course, Noise Hall, Elevator Simulation, and Simulated Day Room) were

Pet therapists are a source of happiness that can uplift the human spirit. Photo courtesy of Dr. Caroline Schaffer and Dr. David Schaffer.

designed to simulate the conditions of the nursing home/hospital where the pet who passed the PUPS Test would be working. The simulations were selected because, although an animal's behavior may be 99% reliable in its own home, the behavior may be extremely different in a novel environment.

The evaluator who marks the test sheets does not determine the pet's admission into the PUPS Program. A panel of three people reviews the pet's body language as reported at all eight stations and then gives the pet either an unconditional pass, a conditional pass, or a failing evaluation.

Every handler/owner is told what strengths and weaknesses were observed through the pet's body language so that the handler will be sensitive to the pet's needs during the visitation. If the pet received a conditional pass, the handler is instructed accordingly and given suggestions that may later enable the pet to earn an unconditional pass. If the pet fails, the handler/owner may seek behavioral counseling so that the pet can be re-tested later.

If the interactions between the pet and the handler are flawed, the panel advises the handler and gives specific recommendations to remedy this relationship. Trust and affection between the cat and its owner/handler appear to be part of what enables a cat to work its magic in unfamiliar surroundings. An owner/handler who is attentive to his pet's needs can prevent unpleasant or dangerous encounters.

It is important to note that the Tuskegee PUPS Test as presented here was designed as a screening test for nursing home and hospital visitations. If a cat were to be invited to a different institution, then the test would be modified to more closely simulate that setting. If, for example, cats will be visiting a school for autistic children, then they would need to be handled by children and subjected to the sounds and manipulations that they might encounter in that setting. Only then can the cats show, through their body language, how they feel about interacting in that environment.

TESTING AND LEARNING NEVER END

Passing a behavior test does not ensure that a cat will react well in the institutional setting. If, in spite of previous screening, he behaves inappropriately at the institution,

Some breeds of cat may be better suited to pet-therapy work than others. American Shorthair photographed by Isabelle Francais.

then he should be removed immediately from the program. It is unkind to the cat and potentially dangerous for those in the institution to force the cat to continue in the program.

Also, just because a cat passed a behavior test does not ensure that he will always be effective. Cats learn quickly to associate unpleasant experiences with particular sights, sounds, and odors at an institution. To prevent "burn out," the cats must receive consistent positive reinforcement at the institution. The owner/handler must remain focused on his cat at all times, reward him for appropriate behavior, and be ready to quickly remove him from any uncomfortable or dangerous situations.

For peak performance, the length of each visit should match each cat's energy level and attention span. Most visitation programs report that one hour is the longest the average cat should be asked to work.

The diverse messages conveyed by a cat's tail illustrate how important a knowledge of feline body language and behavior is for those using cats for therapy. A flicking tail usually indicates ambivalence or a state of emotional conflict, but it also may signal increasing arousal, anger, or annoyance.

For a pet program to be successful, it is necessary that the owner/handler remain with the pet and be responsible for the pet's well-being. It is also advisable for an institutional staff member who understands the patients/residents and who is responsible for their well-being to accompany the pet and the owner/handler.

To continue to be an effective therapist, a cat must rely on his owner/handler to recognize when he is stressed and to act quickly to relieve the stress. Typical signs of stress in cats are those physiological responses associated with stimulation of the autonomic nervous system. Stress may result in overt or subtle behavior changes. Subtleties may range from signs such as dilated pupils, tense muscles, tense facial expression, sweaty paws or cessation of purring or kneading, to slow, tense tail movements or rapid, side-to-side tail movements. Overt signs may include aggression, salivation, panting, urination, defecation, expressing anal sacs, licking lips, grinding teeth, and loud vocalization.

When a cat appears stressed, it is safest to immediately remove him from the eliciting stimulus. For the cat, this means the owner/handler should put him in his pet carrier and either put him in a quiet place in the facility or return him to his home.

Later, the owner/handler should determine whether the cat can learn to accept the stressful stimulus or whether the cat will need to be kept from it. For example, a cat may be fearful of the high-pitched, crackling voice of one particular patient. It may be possible to eliminate the cat's fear or anxiety through a carefully designed counterconditioning and desensitization or habituation program or it may be necessary to stop the visits between the cat and that particular patient.

Sometimes the best decision when a cat is overly stressed is to retire him from the therapy program. Not only will the cat be happier, but the patients/residents may also be happier. Few things are worse emotionally for those being visited by a therapy cat than the perception that the pet dislikes them or is stressed by their presence. Before deciding the appropriate course of action, the owner/handler should consult a veterinarian with training in animal behavior for help with the diagnosis and treatment of the stress-related problem.

PLANNING THE ADOPTION

Anyone hoping to adopt a cat for use in a pet-facilitated therapy program or other animal-assisted therapy program would do well to first study the characteristics of various breeds of cats, learn more about feline behavior, and determine what pet programs exist in the local community. Although it is impossible to know with certainty which kittens will grow up to be good therapists, anyone looking for a prospective therapy cat should select a cat who exhibits a high tolerance to touch.

A kitten who appears shy is unlikely to work well. It is especially helpful to know what happened during a kitten's sensitive period of socialization between two and seven weeks of age. If a cat has good experiences with people and other animals throughout this period, he is more likely to react well later. If, for example, he has little or no experience with children or loud noises when he is two to seven weeks old, he may always be fearful of children and loud noises. Ideally, the prospective therapy cat should have had many positive interactions with many people and animals in many different locations during his socialization period.

Besides being socialized to people and other animals, a therapy cat must be trained to wear a harness. Even if the cat is to be carried, he should wear a harness and leash for added control in case of an emergency. The cat will also need to be trained to travel in a pet carrier. This will give the cat a safe haven should he need to be removed from the people he is visiting. The cat also needs to be trained so that he will not soil the facility or the people he is visiting.

Some cats capable of passing behavior tests may have idiosyncracies that make them inappropriate, even dangerous, for therapy programs. It is the responsibility of the owner/handler to screen out those cats and, if in doubt, seek advice from a veterinarian or other authority on behavior. For example, those cats who tolerate petting for a time—perhaps long enough to pass a behavior test—and then suddenly

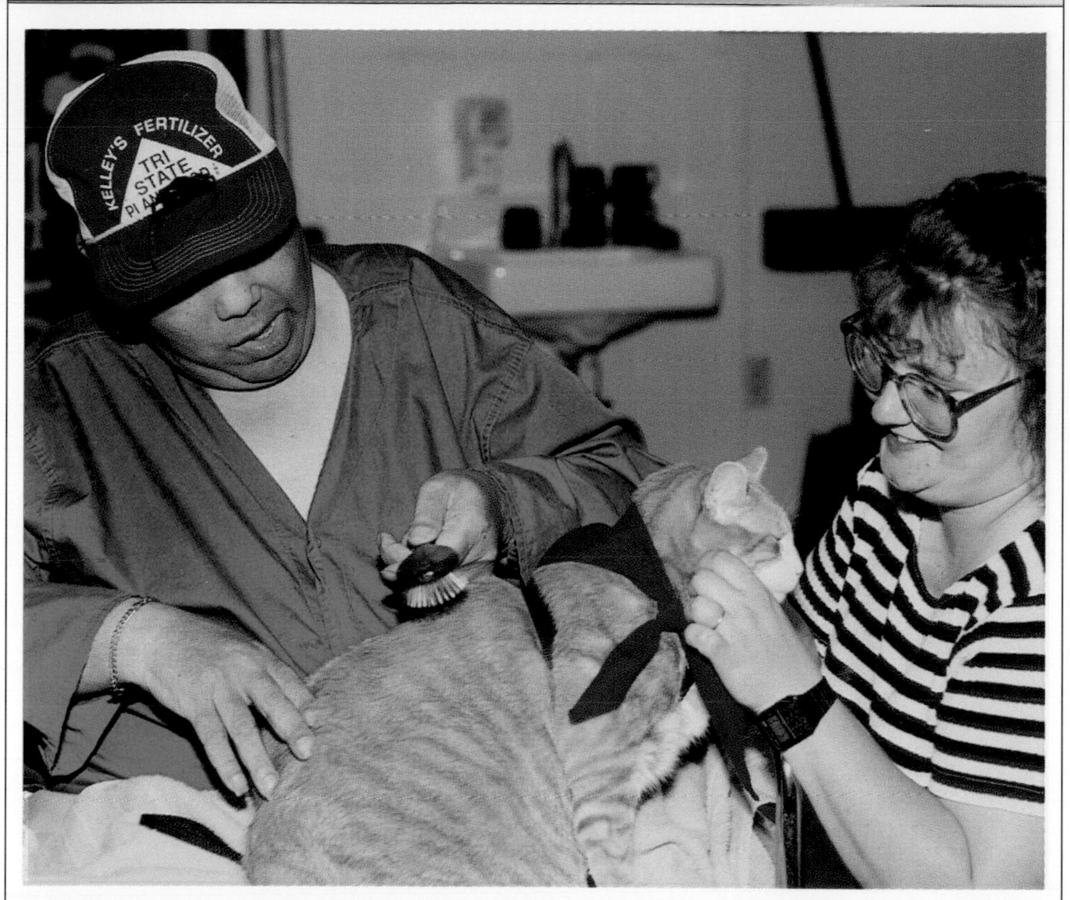

One of the most important requisites for any animal used in pet therapy is its willingness to be touched. Photo courtesy of Dr. Caroline Schaffer and Dr. David Schaffer.

turn and bite the person's hand or wrist would present an unnecessary risk.

SHARING RESPONSIBILITIES

Selecting an appropriate cat is only one step in assuring a successful pet-facilitated therapy program. The institutions, the cats' handlers/owners, and the patients/residents all have their own unique responsibilities if cats are to work their magic. All have a responsibility to be sure the cat feels comfortable and safe.

The responsibilities of the institution include:

1. Providing a safe environment where the cat and his owner/handler will not be injured or become ill,
2. Informing its staff members about the pet program and its goals,
3. Preventing staff members from deliberately sabotaging the program,
4. Alerting its infectious disease control board that cats will be coming so that proper protocol

will be understood and followed,
5. Keeping the cat away from patients/residents who have diseases (such as scabies caused by a notoedric mite) that are known to be contagious to/from cats,
6. Screening patients and staff for allergies and phobias to cats so that they are not forced to be near animals they either dislike or fear,
7. Providing a staff member who will escort the owner/handler and cat during the visitation and who will facilitate the interactions of the cat by looking out for the well-being of the patients/residents, and
8. Carrying liability insurance that will cover the cat and his owner/handler during their service at the facility.

The responsibilities of the cat's owner/handler include:

1. Selecting a cat who will behave appropriately,
2. Making certain the cat is healthy and free of zoonotic diseases,
3. Keeping the cat on a good preventive health program that includes regular vaccinations for rabies, feline distemper, feline leukemia, and feline respiratory infections, tests for internal and external parasites, and yearly physical examinations,
4. Cancelling the visit if his cat is not feeling well on the day of a scheduled visit,
5. Bathing and brushing the cat within 48 hours of each visit and being sure the cat's nails are short and blunt—declawing or using nail caps such as Soft Paws® may be beneficial,
6. Refraining from applying any perfumes or powders that might cause an allergic reaction in the patients/residents,
7. Protecting the cat and reassuring him throughout each visit,
8. Staying focused on the cat and, preferably, keeping a hand on the cat so that he can be removed quickly if an emergency arises,
9. Knowing the cat's behavior well enough to detect when he is getting tired, feeling excessively stressed, or becoming agitated,
10. Knowing how the cat behaves with dogs and other animals whom the cat might encounter during a visit,
11. Providing the cat with an easily accessible escape route,
12. Helping the cat to get comfortable if he is to sit on a person's lap or lie on a bed,
13. Providing the cat with access to fresh, clean water,
14. Taking the cat to a previously approved area to urinate and defecate,
15. Reporting bites and scratches or suspected bites or scratches to the appropriate staff member in the institution and following the predetermined protocol for handling injuries,
16. Notifying housekeeping if the cat urinates, defecates, or vomits and cleaning up after the cat as directed by the institution's guidelines,

Pet therapists, their owners, and the patients they have worked with gather to say good-bye at the end of a visit enjoyed by all. Photo courtesy of Dr. Caroline Schaffer and Dr. David Schaffer.

17. Keeping the cat away from anyone who is eating or who has open food nearby,
18. Prohibiting anyone from feeding the cat during the visit,
19. Keeping the cat from ingesting any medications or chemicals at the institution,
20. Not offering food to the patients/residents,
21. Treating the patients/residents with kindness and respect,
22. Following the rules of the institution,
23. Being prompt, dependable, and appropriately dressed so that his clothing does not get in the cat's way or distract the patient/resident,
24. Accepting the institution for what it is—not trying to change policies, criticize housekeeping, or provide patient care beyond that described under the institution's pet therapy program, and
25. Maintaining confidentiality and respecting the patient/resident's right to privacy.

The responsibilities of the patient/resident include:

1. Being kind to the cat,
2. Accepting that he cannot have a visit if he is abusive toward the cat,

3. Never putting the cat in a potentially dangerous situation,
4. Not monopolizing the cat or prohibiting others from visiting with the cat,
5. Never feeding the cat unless the institution and the owner/handler approve, and
6. Asserting his right to not visit with or handle a cat whom he is fearful of, allergic to, or uncomfortable with.

STOPPING NEEDLESS SEPARATIONS

Significant strides have been made in enabling people in various stages of life to benefit from the human-animal interdependent relationship. Much has changed since Florence Nightingale, the pioneer of modern nursing, mentioned the psychological benefits of pets in her 1893 "Notes on Nursing," and Boris Levinson, child psychologist and father of modern pet psychotherapy, stirred up the psychiatric community with his controversial paper, "The Dog as a 'Co-therapist'," in 1962. Despite the progress, some segments of society are still needlessly separated from animals. More can be done to enable the elderly, the hospitalized, and the immunosuppressed including those infected with the Human Immunodeficiency Virus (HIV) to interact with cats of their own.

Federal laws require that the elderly and handicapped living in federally subsidized housing be allowed to own pets, but, ironically, the elderly who are financially independent and who rent apartments or purchase condominiums in "modern" retirement villages are often prohibited from owning a cat or other pet. How tragic that, while they may have enough money to buy virtually anything they want, they cannot buy a cat! The tragedy is that the cat or another suitable pet would keep them active, add joy to their lonely days and long nights, and provide companionship after their friends and loved ones have died.

The physical and emotional benefits of pet ownership have been proven scientifically to the satisfaction of the U.S. government. Research data on the health benefits of pets for the elderly is convincing. It is time for landlords to change their perception of cat ownership and to welcome tenants who can adequately care for their cats. A few model programs, such as the ones in Massachusetts and New Jersey, have proven that guidelines can be established that will protect the landlord's property and provide for beneficial people-cat interactions.

More hospitals would do well to adopt policies that will allow for patients to have visits from their own cats, not just those who are in pet-therapy programs. Some hospitals have, for example, dedicated one or two rooms for pet visits. Patients can reserve a room and make arrangements for their pet to meet them there for brief visits at assigned times. Actress Elizabeth Taylor and comedienne

A mixed breed cat can make just as good a pet therapist as can a purebred cat. Photo by Robert Pearcy.

Joan Rivers talk openly about smuggling their dogs into the hospital with them. Clearly, visits with a person's own pet have proven to be even more beneficial than visits from pets belonging to volunteers.

The companionship of cats is being denied to many people in another segment of society at a time when they can help the most. Healthy cats with an appropriate temperament can provide physical and psychological benefits to their owners who are immunosuppressed. These benefits usually far outweigh any risk of the owners' getting sick or injured from their pets; nevertheless, well-meaning friends and health care providers frequently recommend that people whose resistance to infection is low due to diseases such as Acquired Immune Deficiency Syndrome (AIDS) or medications such as chemotherapy or corticosteroids get rid of their pets.

Tragically, many people with HIV are losing their pets at the same time they are being shunned by their friends and family. The only time they are touched is when they are poked and prodded in the doctor's office or hospital. For them, a cat may be the only one they can hug and cuddle without fear of rejection. In addition, they may feel that their cat is the only friend who is genuinely happy to see them. Unlike a human friend or volunteer who can visit only a short while, a cat will be absolutely delighted to spend the entire day and night with his owner.

Although it is true that several agents known to infect both people and cats have been found in HIV-infected people, medical scientists believe that people infected with HIV rarely get these agents directly from a pet. More often, they are acquired from contaminated soil, food, water, wild birds, or infected people.

An immunosuppressed person can care for his cat in such a way that will minimize his risk of contracting a zoonotic infection. Veterinarians can provide an expanded preventive health care program designed to match the client's level of immunosuppression. The client can also follow the guidelines described in "HIV/AIDS and Pet Ownership," a brochure available through the Tuskegee University School of Veterinary Medicine, Tuskegee, Alabama 36088, to ensure a safe and healthy relationship with his cat. As has been shown in other areas of pet-facilitated therapy and animal-assisted activities, the risk of transmission of diseases between animals and people can be minimized by having the proper information.

SUMMARY

Cats and other companion animals throughout the nation and the world are putting sunshine into the hearts of people who are isolated because of either emotional, physical, or behavioral problems. Undoubtedly more cats will be allowed to work their magic as more veterinarians, physicians, other health care providers, and animal lovers show that not all cats are independent, wild, and unpredictable. By passing comprehensive behavior tests and health examinations and by being paired with owners/handlers they can trust, cats are proving to be invaluable members of today's health care team.

ADDITIONAL READING:

American Veterinary Medical Association: Guidelines for Animal-Facilitated Therapy Programs, American Veterinary Medical Association Directory, Schaumburg, IL, 1994.

Anderson, C.: *Pets in People Places: Responsible Pet Ownership in Multi-Unit Housing,* Massachusetts Society for the Prevention of Cruelty to Animals, Boston, MA, 1994.

Arkow, P.: *Pet Therapy: A Study and Resource Guide for the Use of Companion Animals in Selected Therapies,* The Humane Society of the Pikes Peak Region, Colorado Springs, CO, 7th Edition, January 1992.

Beaver, B.: Feline Behavior: *A Guide for Veterinarians,* W. B. Saunders Company, Philadelphia, PA 1992.

Beck, A.; Katcher, A.: *Between Pets and People, the Importance of Animal Companionship.* Pedigree Books, Putnam Publishing Group, New York, 1983.

Bergler, R.: *Man and Cat: The Benefits of Cat Ownership,* Blackwell Scientific Publications, Oxford, 1989.

Cusack, O.: *Pets and Mental Health,* The Haworth Press, New York, 1988.

Delta Society. *Pet Partners Volunteer Training Manual,* Renton, WA, 1993.

Dorsey, G.: Schaffer, C.; Ferguson, J.A.: *"HIV/AIDS and Pet Ownership,"* brochure, Tuskegee University School of Veterinary Medicine, Tuskegee, AL 36088, July 1994.

Fogle, B.: *Know Your Cat—An Owner's Guide to Cat Behavior,* Dorling Kindersley, Inc., New York, 1991.

Rfau, H.: *PAT at Huntington, A Volunteer Program of Pet-Assisted Therapy,* Training Manual, Huntington Memorial Hospital, Pasadena, CA, 1990.

Schaffer, C.: Phillips, J.: *The Tuskegee PUPS Test for Selecting Therapy Dogs* (video), distributed by Tuskegee University School of Veterinary Medicine, Tuskegee, AL, 1993.

Schneck, N.; Caravan, J.: *You're Ok, Your Cat's Ok,* Chartwell Books, Inc., Secaucus, NJ, 1993.

University of Tennessee: *HABIT (Human-Animal Bond) in Tennessee: An Overview* (video), distributed by University of Tennessee, Knoxville, TN, and by The Delta Society, Renton, WA, 1989.

Dr. Ernest Rogers obtained his Bachelor of Arts in psychology with emphasis on physiological psychology. He also obtained a Bachelor of Science in biology. Both baccalaureate degrees were obtained at Guelph University, Guelph, Ontario, Canada. In 1991, he graduated with a Doctor of Veterinary Medicine from Tuskegee University, Tuskegee, Alabama. Currently, Dr. Rogers is obtaining his Doctor of Philosophy, with an emphasis in veterinary pharmacology, from Virginia Polytechnical Institute and State University, Blacksburg, Virginia.

The Brain, Biology, and Behavior

By Ernest Rogers, BA, Bsc, DVM
Virginia-Maryland Regional College of Veterinary Medicine
Virginia Polytechnical Institute
Duckpond Road
Blacksburg, VA 24062-0442

INTRODUCTION

Many individuals, including veterinarians, view behavior solely as a manifestation of learning, socialization and psychology. However, behavior is a final common output of many body systems in concert with previous learning and the environment. Body systems, including the endocrine, cardiovascular and neurological systems (and others) influence the type, display and intensity of behavior. This chapter elucidates the need for a comprehensive view of the brain and body systems as well as for learning and the environment.

Veterinary clinical behavior counseling often deals with the aspects of psychology that have a direct impact on information acquisition—learning! In many instances, behavior is most easily modified, problems are corrected or avoided, and training to a goal most easily accomplished by using learning principles. In fact, most of this book is dedicated to this premise. Although learning principles are very successfully used in most instances, other problems appear to be more resistant to a purely behavioral modification approach. There are a number of reasons for this. First, the principles applied may be incorrect or inappropriate for the situation, the patient or the environment. Second, the problem may be dynamic, complex or deep-seated. This may require medical or drug therapy in addition to behavior modification. Third, the problem may be based on a physiological condition or disease that has little to do with behavior modification. For example, a dog with epilepsy is unlikely to respond to training exercises. This chapter is intended to serve as an introduction to the complex organization and interaction between the biology of the brain and behavior.

THE BRAIN

The brain is located within the protective bones of the skull. This organ is of a "jelly-like" consistency. The surface of the brain varies in complexity depending on the species.

The brain is organized, grossly, by surface anatomical features. In general, distinct lobes can be visualized from an overall surface view. These features allow some distinction as to functional areas, as they relate to the whole brain. The main organizational and "thinking" area of the brain is centered in the frontal lobe. The temporal lobe is associated with the emotional and auditory (hearing) functions. The occipital area is considered the seat of the ability to interpret visual images (sight). The cerebellum acts in concert with the parietal lobe to activate and coordinate movement. The pituitary lobe is the link between the brain and the endocrine system (hormones). The pituitary is found at the base of the brain. This lobe has two distinct parts, the anterior (front) area and the adjacent posterior (back) pituitary area. Both areas are responsible for the release of separate unique hormones that influence other endocrine organs (i.e., pancreas, adrenals and sex organs). This ultimately causes increases or decreases in hormonal levels.

The brain is organized internally into four functionally distinct but interactive levels. The first level of organization is that of the cell unit. The basic cell of the neurological system (brain, spinal

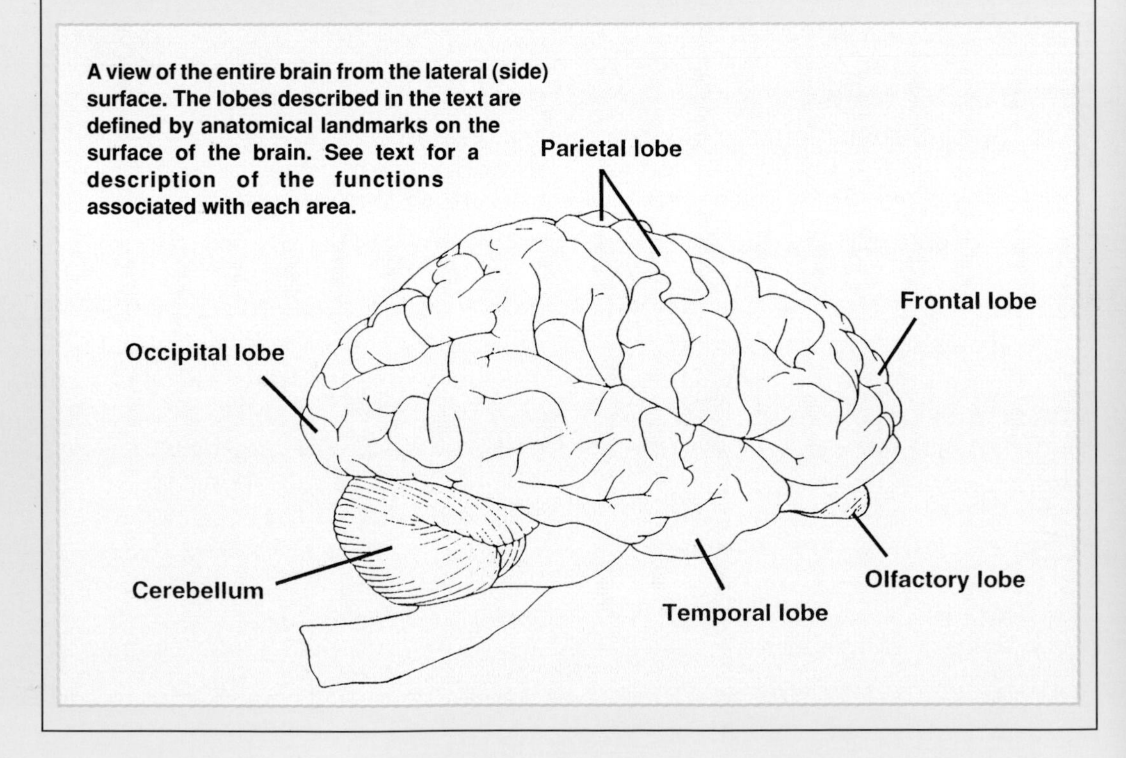

A view of the entire brain from the lateral (side) surface. The lobes described in the text are defined by anatomical landmarks on the surface of the brain. See text for a description of the functions associated with each area.

cord and peripheral nerves) is called the neuron. Neurons consist of a cell body with two distinct ends, one of which forms a long "wire-like" process (the axon). This cell, at its most primitive level, is a communication cell, carrying information from one location to another. The neuron accepts bio-electrical information at one end (dendrite), transfers this information across its cell body, and delivers the information to the next cell through its axon. The distribution of information is therefore from the axon of one cell to the dendrite of the next cell. Information is transferred across a minuscule space (synapse or synaptic gap) between cells by chemicals released by the end of the axon. It is the amount and timing of the chemicals released that conveys information to the dendrite. The chemicals which convey the information by crossing the synapse between cells are termed neurotransmitters. Examples of neurotransmitters that are of particular interest to behaviorists are: serotonin, gamma-amino-butyric-acid (GABA) and dopamine, to name a few.

The second level of organization within the brain is the grouping of "information-carrying neurons" into complexes of cables or groups of cell bodies. White matter of the brain is an accumulation of neuronal axons traveling together from their cell bodies to another location. The gray matter of the brain consists of groups of cell bodies located in one common area. Groups of neuronal cell bodies form a nucleus which is a discrete area with specific functions and unique inputs and outputs.

The third level of organization is that of combinations and interactions between larger groupings of nuclei that tend to coordinate and orchestrate specific behaviors. Often, both the second and third levels of organization are redundant, with mirror images of similarly functioning groups of nuclei on both the right and left side of the brain. Examples of grouping of nuclei include the amygdala, hypothalamus and the hippocampus.

A fourth level of organization relates to the complex interactions of the various groupings of areas of the brain that result in a coordinated directed pattern of behavior. This serves the behavioral and survival needs of the animal. The preceding three levels of the brain must be functioning properly to assure that the final common output, behavior, is appropriate and of the correct intensity for the situation. An example of this level of organization is the limbic system, an important area for emotional behavior in both man and animal.

The brain may become dysfunctional at any level of organization. The disturbance may result in bizarre, inappropriate or inconsistent

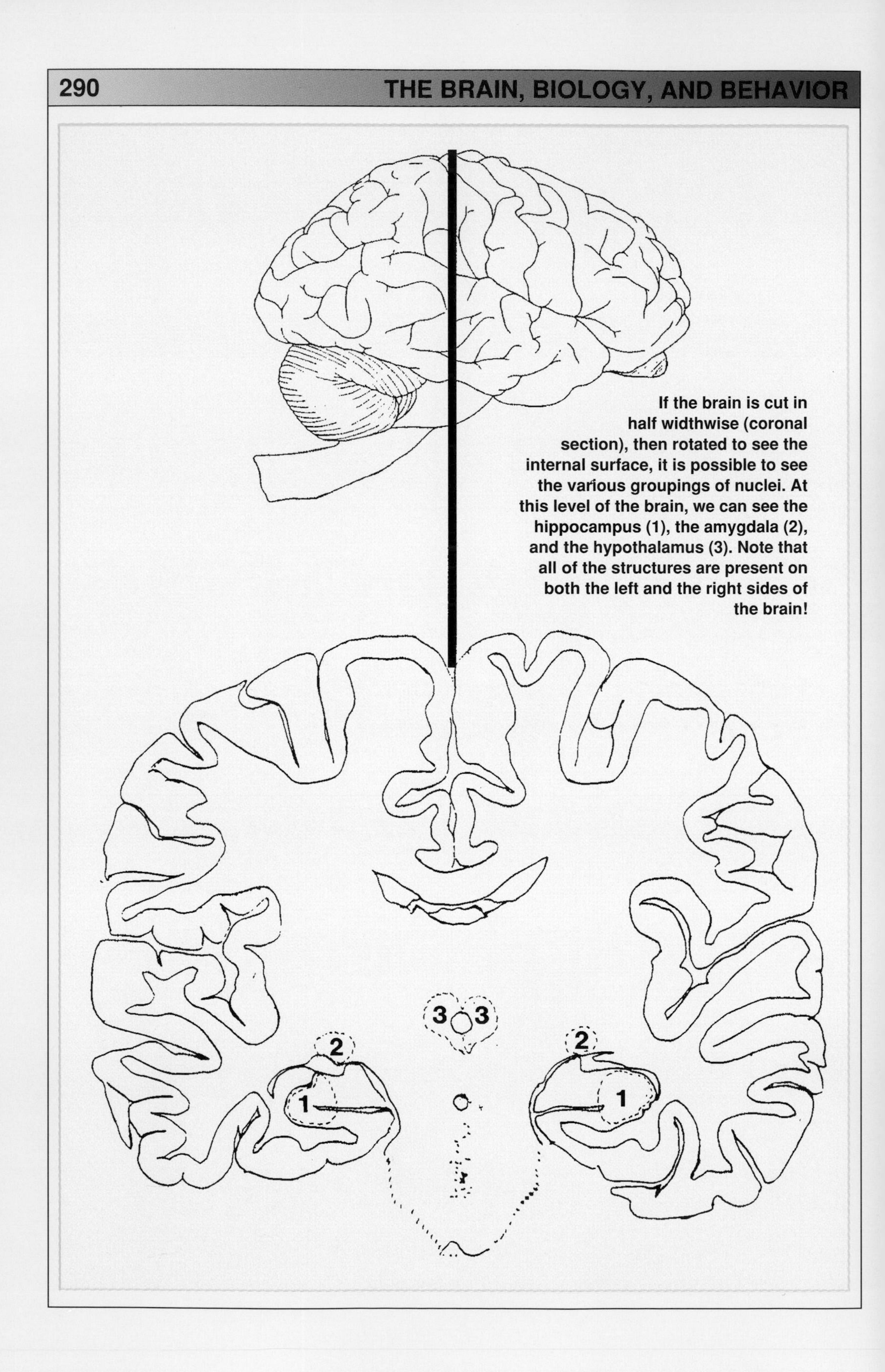
If the brain is cut in half widthwise (coronal section), then rotated to see the internal surface, it is possible to see the various groupings of nuclei. At this level of the brain, we can see the hippocampus (1), the amygdala (2), and the hypothalamus (3). Note that all of the structures are present on both the left and the right sides of the brain!
3
3
2
2
1
1

behavior. Hormones can alter the biology and chemistry of the body (and brain) and therefore the expression of behavior. Further, alterations in body chemistry tend to result in behavioral changes. In part, this may explain some strange behaviors (such as pacing and weaving) seen in captive and highly stressed animals. Let's look at some specific behavior problems and how they may be related to brain function or dysfunction.

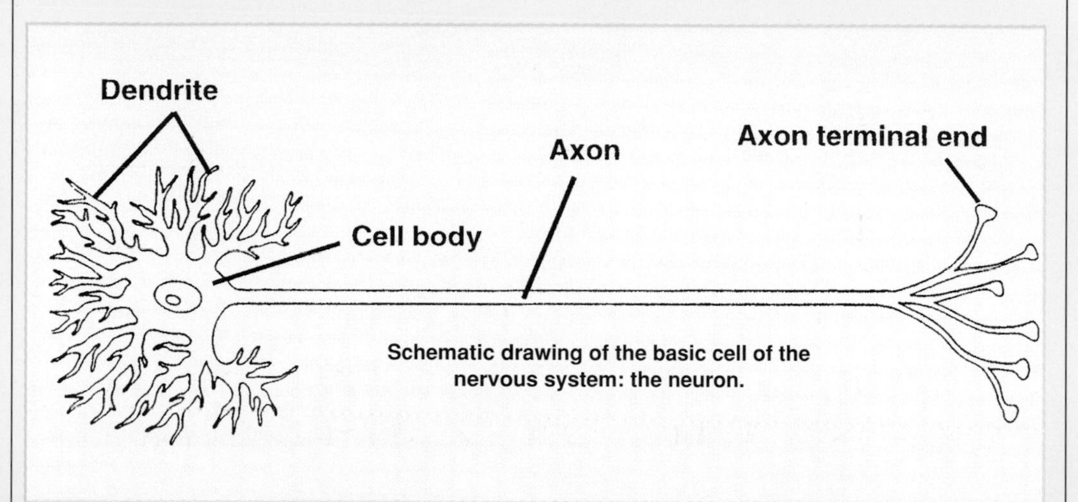

Schematic drawing of the basic cell of the nervous system: the neuron.

AGGRESSION

Though aggression may be learned or instinctual, there is a wealth of scientific information indicating that the expression of aggression, the intensity of the aggressive behavior, and the initiating events that lead to aggression are associated with discrete areas of the brain.

The part of the brain most commonly thought to be responsible for emotional behaviors (including aggression) in animals is the limbic system. The limbic system is buried deep in the temporal lobe of both sides of the brain. This system consists of several nuclei that, experimentally, have been associated with aggression.

Damage to the septum (a part of the limbic system) increases the likelihood and severity of attack behavior in animals. This has been termed "septal rage syndrome" and usually consists of biting and scratching attacks directed at some objects. Induced damage to another part of the limbic system (the ventral nucleus of the hypothalamus) can result in a similar increase in attack behavior. In contrast, another arca in the limbic system, the amygdala, is associated with decreased aggression.

Aggression may also be modified by arousal (a function of the part of the brain called

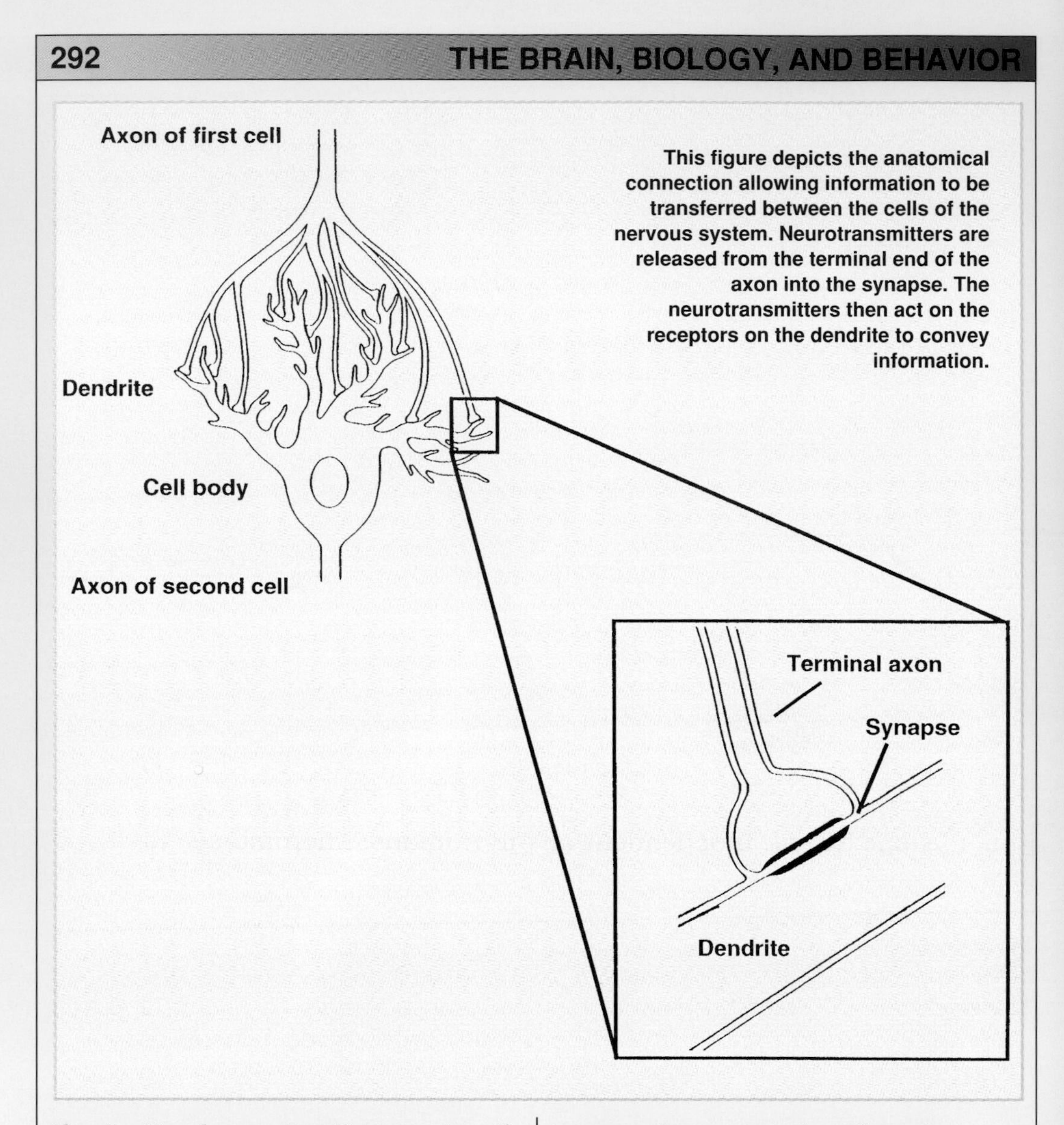

This figure depicts the anatomical connection allowing information to be transferred between the cells of the nervous system. Neurotransmitters are released from the terminal end of the axon into the synapse. The neurotransmitters then act on the receptors on the dendrite to convey information.

the "reticular activating system") and hormonal levels. These can indirectly lead to increased or decreased aggression. Intact males, in general, tend to be more aggressive and more protective (or territorial) than their neutered counterparts. The hormone which appears to modulate this aggression is testosterone. It is for this reason that some practitioners suggest castration as a solution to an aggressive unneutered male. Similarly, some female animals have demonstrated increased aggression during periods of estrus (heat), when compared to those in non-estrous periods. Though these hormones predominantly originate from the reproductive organs, the brain modifies the behaviors with respect to the environment, the species involved and the specific situation.

Thus, brain behavior must be considered whenever dealing with an aggressive animal. Behavior modification alone may not be sufficient to correct the situation.

FEEDING BEHAVIOR

Feeding behavior, in part, is controlled by specific regions in the hypothalamus. A "satiation center" can be stimulated to reduce hunger and decrease appetite in an otherwise hungry animal. In a similar fashion, some nuclei of the lateral nucleus of the hypothalamus (hunger center) may be stimulated to increase hunger and feeding behaviors. Hormones also play a role. Insulin and glucagon, as well as the nutrient glucose, can cause fluctuations in hunger. This can be, to some extent, independent of brain activity.

Pica is a term used to describe the consumption of non-nutritive, non-beneficial, non-food items. For example, the animal that eats sticks or stones receives no benefits for this action. This abnormal behavior may be mediated by a neurologically related need for oral gratification. If so, successful treatment would require more than just behavioral modification and training.

SEXUAL BEHAVIOR

Sexual behavior has several ties to brain function. Experimental evidence exists to show that lesions in the preoptic region or anterior hypothalamus result in a decrease in male sexual behavior. In contrast, removal of areas within the temporal lobes (primarily the amygdala) has been associated with significantly increased libido in both.

Pheromones are specific odor-related chemicals given off by animals at specific times during their reproductive cycle. These pheromones communicate information about the receptivity of the female (readiness for mating) or the proximity of a male. These chemicals may act to increase the libido and sexual behavior in both males and females. Olfactory lobes, which are the part of the brain responsible for the sense of smell, are much larger in dogs and cats than the comparative structures in humans. Pheromones, and smell in general, are of significantly more importance to animals than to humans.

Hormones are released and act within the body in response to light cycle, pheromones, and ambient temperature, all of which are monitored by the brain. Hormones are known to change the sexual drive and the associated behaviors. The hormone affecting the sexual drive in males is testosterone. In females, luteinizing hormone, estrogen, and progesterone appear to play an active role.

By using drugs (i.e., female hormones in a sexually active male) some behaviors associated with sexual maturity may be reduced. We may also modify sexual development and associated behaviors by castration and ovariohysterectomy (removal of the ovaries and uterus).

Maternal behavior is related to changes in the brain as a result of hormonal fluctuations. Maternal interactions are a result of a combination of changes in brain chemistry (brain environment and neurotransmitters) and body chemistry (hormones). The hormones most notably associated with this behavior are progesterone, estrogen and prolactin. Once these hormones take effect, the individual animal becomes more nurturing. Some typical maternal behaviors include nesting, nursing and increased attentiveness. In the case of pseudopregnancy, all typical maternal behaviors may be seen in the presence of hormonal influences without an actual pregnancy. This demonstrates the powerful influence hormones may have on an animal's behavior.

Though no single aspect of sexual behavior is more important than any other, drug therapy using sex hormones can be useful in modifying some inappropriate behaviors.

COMMUNICATION, OLFACTION, AND URINATION AND DEFECATION

Urination, in some circumstances, can be a form of communication in both dogs and cats! This type of communication may tell of one animal's territory or the fact that a female is in heat. As mentioned previously, the olfactory lobes of animals are much larger and more active than those of the human.

The importance of urination or defecation as a form of communication for dogs and cats cannot be overlooked. In most cases, urination and defecation are simply natural body functions. In some cases however, body position, location and quantity of urine or feces can be a form of communication. Inappropriate defecation or urination must therefore be examined as an inappropriate communication, as a break in housetraining or possibly as a medical problem.

Hormones may decrease sexual communication and territoriality by their actions on the hypothalamus. Sedatives may decrease arousal associated with marking behavior. Surgically altering an animal's sense of smell (surgical olfactory agnosia) may alleviate problems (such as feline urine spraying) that are unresponsive to other therapies.

OBSESSIVE-COMPULSIVE BEHAVIOR IN ANIMALS

Obsessive-compulsive (OC) behaviors are those behaviors that are disruptive to the animal's normal ability to function. These behaviors may fall into two general categories. Some are repetitive, unvarying behaviors that fail to have an apparent function for the animal. An example is self mutilation (i.e., excessive licking or chewing) in both dogs and cats.

A mixed breed kitten in a state of watchful attentiveness. Photo by Robert Pearcy.

The pituitary gland (hypophysis) is a pendulous gland found suspended beneath the base of the brain at the hypothalamus. In fact, to see this view, the brain was cut in half lengthwise (saggital section) to expose the pituitary. The pituitary gland is composed of two adjacent portions. One is the anterior pituitary (adenohypophysis); the other is the posterior pituitary (neurohypophysis). Each is responsible for the release of different hormones. The release of hormones is dictated primarily by the hypothalamus.

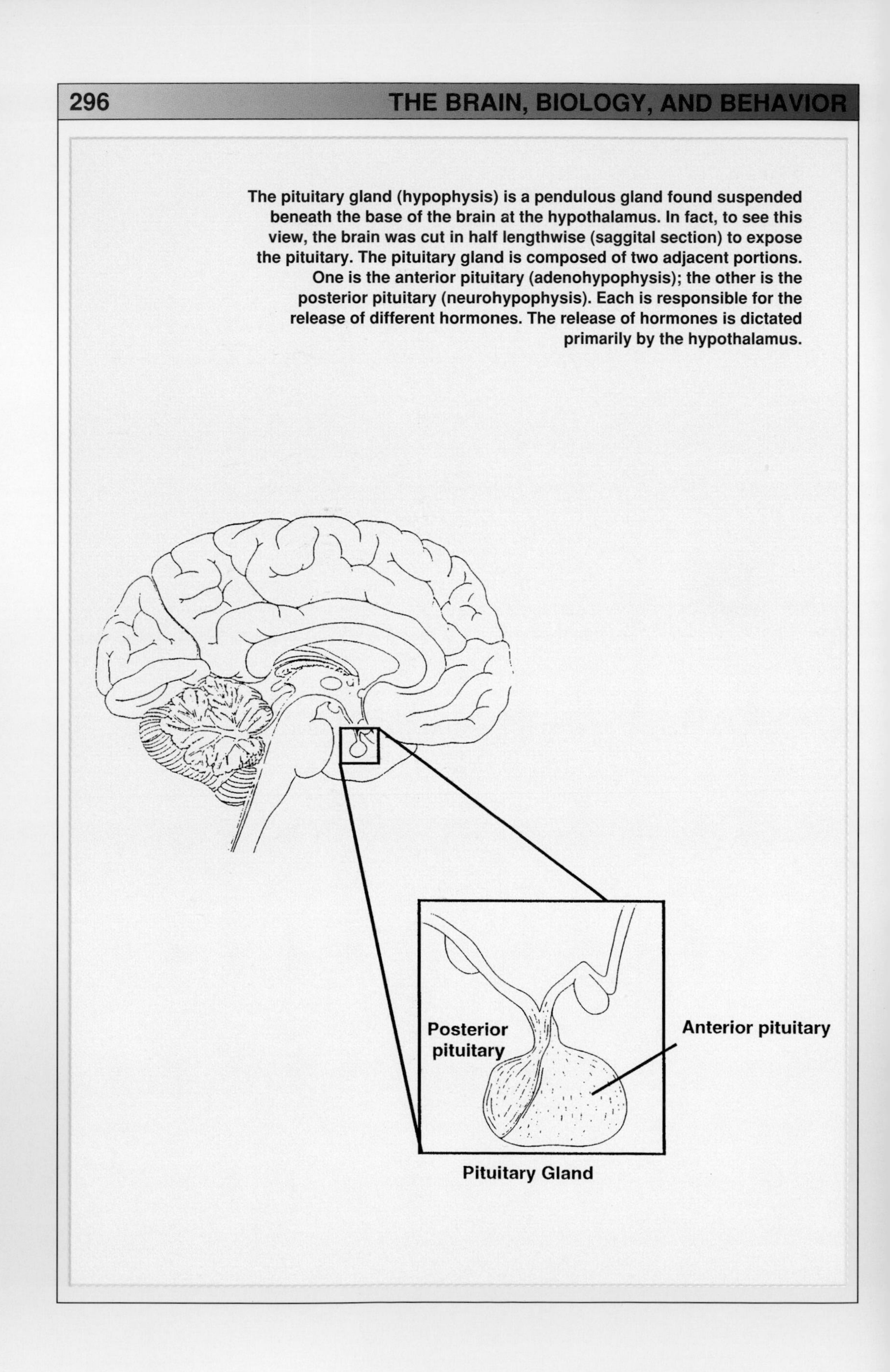

Pituitary Gland

A second category of obsessive-compulsive behaviors is not associated with a repetition of movements. Examples include flank sucking in dogs, standing motionless (freezing) in cats, or staring into space in both dogs and cats.

The signs of OC behavior are as varied as the behaviors themselves. It is likely that multiple brain systems are involved in the development and establishment of these behaviors. Visual hallucinations, as in staring behavior, may be related to dysfunction in the visual cortex (occipital brain).

Experimental evidence has been accumulated to indicate that some compulsive behaviors are associated with a disturbance in neurotransmitters scattered throughout the brain. The neurotransmitters most often suggested are serotonin and dopamine. Though much work is yet to be done in this area, there is thus an exciting potential for drug therapy to effectively manage these problems. OC behaviors demonstrate our limited understanding of the brain and its functions as they relate to behavior.

SUMMARY

The examples given above are only a few of the many that demonstrate the complexity of interactions between the brain, biology, learning and behavior. This illustrates the need for a medical and scientific approach to behavioral diagnosis and treatment.

The understanding of the discrete areas of the brain and their effects on behavior is a good beginning, but it must must be tempered by the fact that multiple areas of the brain may perform the same functions. One must be aware that, even armed with the most current knowledge, some treatments will fail due to our inability to fully understand the interaction between the brain, biology, and behavior. Further research and clinical experimentation will lead to a better understanding of the brain and associated biology. This is the key to successful diagnosis and treatment.

ADDITIONAL READING

Carlson, N.R., *Physiology of Behavior,* Allyn and Bacon, Inc. Boston, MA, 1977.

Hart, B.L., *Feline behavior*, Veterinary Practice Publishing Company, Culver City, CA, 1978.

Kandel, E.R. ; Schwartz, J.H.; Jessell, T.M., *Principles of Neural Science* (3rd Edition), Appleton and Lange, East Norwalk, CT, 1991.

Marder, A.R., Voith, V. (editors), *The Veterinary Clinics of North America* (Small Animal Practice) "Advances in Small Animal Behavior," W.B. Saunders, Philadelphia, PA, 1991.

Walker, E.L., *Conditioning and Instrumental Learning*, Brooks/ Cole Publishing Company, Belmont CA, 1969.

Dr. Robert Holmes graduated as a veterinarian from the University of Edinburgh (1969) and received his PhD in animal behavior from the University of Bristol (1975). He spent 16 years teaching, researching and consulting in a wide range of domestic animal behavior and welfare matters at the New Zealand Veterinary School (Massey University). In 1990, he started a full-time animal behavior consulting service throughout Australia. Cats make up about 20% of the cases. Dr. Holmes was the first Member and so far is the only Fellow of the Australian College of Veterinary Scientists in animal behavior. He shares his household and family attention with two Abyssinians.

Outdoor Cats-Indoor Pets

By Robert J. Holmes, BVM&S, PhD, MRCVS, FACVSc
Animal Behaviour Clinic
PO Box 135
Malvern Vie 3 144
Australia

INTRODUCTION

Cats can be happily kept inside all the time. Many people do so and would have it no other way. They say they have deeper and more satisfying relationships with their cats and that those cats are healthier and live longer. While living happily insidc, cats are not getting hit by cars, being injured in cat fights, catching infections such as feline leukemia virus and feline immunodeficiency virus (Feline "AIDS"), being stolen, hunting and possibly killing wildlife, urinating and defecating on neighbors' properties, and harassing or being harassed by other animals. Clearly there are many good reasons for permanently keeping cats indoors.

ISN'T IT CRUEL?

Many people feel that it is crucl to confinc cats becausc they think of them as "free spirits" that should be allowed to roam at will because of their nature. They seem to give little thought to the possible consequences listed above. So, how can we resolve this dilemma? We can do so by enriching the daily life of the indoor cat to replace some of the stimulation and activity it would otherwise receive as a free-roaming animal. This environmental enrichment puts complexity, unpredictability and choices into a cat's daily life. Without these things, many animals and people become frustrated in confinement and show signs of boredom—greater reactivity, irritability and exaggerated or unusual behavior.

HOW IS IT DONE?

Environmental enrichment aims to satisfy a cat's need for interaction with its environment. This can be done in many ways, some of which suit some cats better than others. Cats are notoriously individualistic. Some activities involve the owner in active participation, while others just have to be set up and left for the cat to use when it wishes. By doing more for their cats, owners also enrich their own lives.

These suggestions are for normal healthy cats. You should discuss them with your veterinarian to make sure they are appropriate for your specific cat.

Chasing and Jumping

Small fast-moving objects cause the innate chase response in kittens. Most mature cats will continue to show it, particularly when they have practiced it all their lives. This can be done with small balls, such as practice golf balls that are hollow and have holes in the surface, or items such as scrunched up pieces of newspaper, pulled quickly and erratically on the end of a string. Some people even tie the objects onto fishing lines and poles so that they can cast out and move the object over a bigger area without the cat seeing them do so. Furry, feathery or flapping things are particularly attractive to cats. Patches of bright light, such as the reflection from a watch face or mirror, often get cats chasing. A hand-held laser pointer that gives a brilliant red spot under any household conditions is a very convenient way of exercising cats. Some cats, particularly the younger ones, will jump and strike at soap bubbles, which should be made from non-toxic soap. Quite a few owners admit to playing

A blue Abyssinian surveys the scene from a high-level walkway at the top of a seven-foot-high scratching post. Photo courtesy of Dr. Robert Holmes.

This Persian is about to sink its claws into an armchair. Some cats will scratch household furnishings only when their owners are not present. Photo by Robert Pearcy.

and really enjoying hide-and-seek with their cats.

Feeding

We can make feeding more natural by getting our cats to search for food and by providing it in a form that needs chewing. If you feed dry food, you could put it in small clumps on the floor progressively farther away from the bowl each day. The clumps can eventually be scattered throughout the house in different places each day so that your cat has to search them out. The food in the bowl can be made harder to eat. Whole raw chicken wings*, corn cobs* and a large cube of tough meat require more chewing than dried or canned food. This also improves dental hygiene by rubbing off some of the plaque that builds up all the time on the teeth and can result in bad breath and tooth loss.

Cats like a change to their diet—as long as it is highly palatable! They have even been known to eat things we would expect to be less palatable such as bread, pasta, raw vegetables and curry. It is amazing what they will eat when it comes from the hand of someone they like. You can also take advantage of this to do some training.

Obedience Work

It might sound a bit radical, but cats can be obedience trained with the same principles of positive reinforcement as dogs. Why do you think they suddenly appear when the

*Personal recommendation of author. Not advocated by most veterinarians.

refrigerator door is being opened? That is not innate behavior; it has been learned. It's amazing what will be learned when you are hungry and your behavior results in food being given. Cats can easily be taught to come, sit, stay, lie down, and retrieve. Reward the desired behavior immediately as it occurs. Break down the learning task into small steps and start at the beginning. Train with very small pieces of the most palatable food. In this way, your cat will just get a taste and not a stomach full, which will satisfy its hunger. Once you have taught several commands, they can be randomized in order and times of day they are given. Such a training session, particularly when it entails working on a new command, will add complexity, unpredictability and choice to your cat's daily life.

Watching an Interesting Scene

Given the choice, cats will vote with their feet and show us that they like to watch a changing scene. They will choose to sit or lie for long periods in safe places where they can watch the world go by, whether it is street activity, people or animals. With a little bit of thought, we can usually provide that safe and interesting area.

High-Level Walkways

One way cats can get to a vantage point is by jumping or climbing. You can make this

A cat and his young owner enjoy a quiet moment together. Photo by Robert Pearcy.

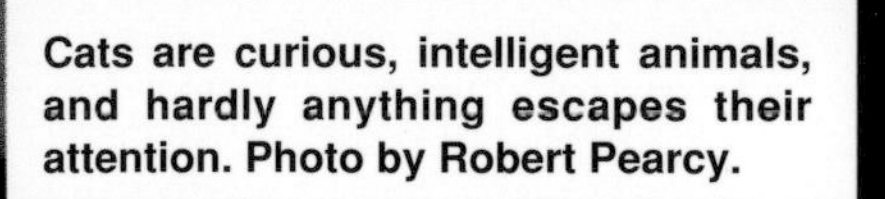

Cats are curious, intelligent animals, and hardly anything escapes their attention. Photo by Robert Pearcy.

This sturdy-looking scratching post offers two different surfaces: carpeting and rope. Some cat owners provide their cats with several scratching posts, located at strategic positions around the house.

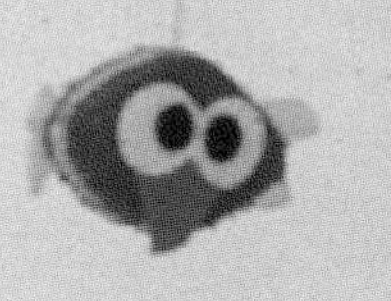

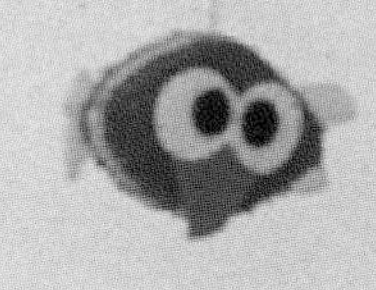

easier and encourage them to use the height of the rooms by providing walkways between high points. Shelves can be strategically placed on walls, or narrow pieces of timber can be placed between beams.

Indoor "Tree"

A convenient way of cats getting access to high points is up a tall scratching post that they will climb as though it were a tree. If the cat cannot climb, for instance if it has no front claws, then a series of shelves could be embedded in a tall post. The cat can then climb by jumping from shelf to shelf.

Get-Away Areas

Given the chance, many cats will lie for long periods in small high places from where they can watch the activity below and presumably feel secure. It is a good idea to provide access to such areas for anxious cats and where there is more than one cat in the house. This can be easily done by closing the lid of a cardboard box of suitable size (about 14"x12"x10" for an average sized cat). Turn it upside down and cut a hole in the middle of one end just big enough for the cat to get in and out. Put in an unwashed garment, such as an old sweatshirt of its favorite person, and place in the highest accessible place in the house. As they are so cheap and quick to make, you can experiment with several of them in different places. High-level walkways, very tall scratching posts or

With all of the kinds of cat foods that are available, it is easy to provide your cat with a diet that is nutritious as well as satisfying. Photo by Robert Pearcy.

indoor "trees" can give access to these places.

Scratching Post

Cats can be trained to use a scratching post and not to use other surfaces for their stretching and scratching exercises. Cut pile carpet is an

Cats have to scratch to remove the dead cells from their claws.

attractive surface through which they can drag their claws. However, a material that can be torn out is preferred. This may be a loosely woven material or a soft wood composition board. The scratchable surface could be firmly attached to a post at least two feet high that is firmly held in position, usually by a heavier base. The forefeet of kittens can be gently placed up the post and drawn down it. By rewarding the kitten with praise and stroking while it is scratching and food when it has finished, it usually quickly learns to exclusively use an attractive post.

Shouting at or spraying a cat with a water pistol is likely to reduce scratching in your presence. However scratching may well still be done in your absence. This can be diverted by temporarily putting a scratching post in front of the scratched surface that is protected by a non-scratchable cover such as wood, steel or thick plastic. Reward the cat for using the post. When it is using it consistently, then move it less than a foot each day toward an acceptable position. As cats tend to stretch and scratch after a rest, the post is best placed close to the cat's sleeping area. You may find it helps to have a scratching post in each room. Once the cat is using the post in the new position, the protection over the scratched area can be removed. If the cat goes back to scratch the area you find undesirable, it means that surface is more attractive than the post. The post could be made more attractive and/or mousetraps could be hung with their bottoms facing out on the surface you don't want the cat to scratch. When the cat touches the back of the trap, it springs out from the scratched surface

Atop a secure perch, this mixed breed cat watches the world go by. Photo by Robert Pearcy.

and cannot snap shut on the cat's paw. There are soft plastic "paddle" attachments commercially available for mousetraps to reduce the chances of a cat getting hurt. They also increase the visual impact of the trap going off .

The effects of scratching can be reduced by regular trimming of the cat's nails or by gluing rounded plastic tips over the ends of the nails (e.g., SoftPaws™).

There are different attitudes about declawing cats to stop scratching problems. While it is still not uncommonly done in the USA and Canada, it is very rarely seen in Australia, New Zealand and Great Britain. The present policy of the Australian Veterinary Association is that the removal of claws, particularly those that are weight bearing, to prevent damage to furnishings is not acceptable unless the only other option is euthanasia.

Paper Bags and Boxes

They say that "Curiosity killed the cat," and watching cats check out newly arrived containers shows how keen they are at investigating. Allowing them access to these new shapes and smells will add novelty to their lives.

Entertainment Box

Taking advantage of their well-known tendency to investigate things with their paws, we can

A Birman playfully swiping at its toy. Offering your cat a variety of play objects will help to keep it from getting bored. Photo by Isabelle Francais.

American Shorthair, silver tabby. Photo by Robert Pearcy.

put small objects inside a box in which there are holes through which the cat can put its paws but through which it would be very difficult to remove the objects. Such entertainment centers are also commercially available.

Cats enjoy toys that contain catnip or other types of cat grass. Photo courtesy of Cosmic Pet Products.

Catnip, Cat Mint and Cat Grasses

These plants can be successfully grown indoors in pots from seeds or small plants that are commercially available. Many cats will visit a catnip plant each day to sniff, rub, grasp, roll alongside and kick at it. This seems to be play and can be observed by both sexes of reproductive age, whether or not they have been neutered. Catmint and cat grasses are attractive to many cats and are more likely to be chewed than some of your other indoor plants. This gives the cats fresh vegetation to eat, which they would otherwise do outdoors.

Trips Outside

Most cats enjoy a trip outside whether it is on a lead and collar or harness, in their owners' arms, or in cars. They can be trained to walk on a lead by reinforcing the walking forward with tiny pieces of favorite food. The differing sights, sounds and smells add to daily variation in stimulation.

Outdoor Enclosures

Various structures can be used to allow cats out into fresh air but restricting their movements to certain areas. Wire netting can be used to enclose an area alongside the house just like an aviary for birds. Enclosures of different sizes can be used in different sites with tunnels between them and the house. A modular system allows for expansion to a wide range of circumstances.

Companion Cat

For cats that are left on their own for long periods each day, it is a good idea to provide a feline companion. Sociable interaction will enrich their daily lives. The younger they are introduced, the greater the chances of getting along amicably most of the time. There may still be fights and chases that are not playful but

Collars and other cat accessories come in a number of attractive styles. Photo courtesy of Hagen.

seem to be part of normal living. Getting littermates gives you the best chances of a pair getting along. Where other cats are to be introduced, it is preferable to do so when they are kittens, and to have them arrive at the property at the same time. Urine-spraying and fighting are less likely when all the cats are spayed females as compared to having one or more neutered males in the house. Bringing older cats together, particularly when one has been resident for some time, may lead to hissing and fighting, defecation and urination out of the litter tray, urine-spraying and one or more cats becoming reclusive. Tolerance can increase with time and by using such methods as: feeding them progressively closer and closer together; rubbing them alternately with the same unwashed towel to transfer their smells between each other; and with drugs.

Japanese Bobtails. Cats that are kept indoors can live perfectly happy lives. Photo by Robert Pearcy.

WHAT ARE THE PROBLEMS?

Clearly, problems can occur with keeping cats confined all the time. Cats that are suddenly confined to their home may pace and vocalize at the time and place they usually go out. This is the time when they would benefit from chasing play. Scratching furniture and intolerance between cats has been discussed. Jumping or walking on surfaces can be prevented by strategically placed mousetraps or a low voltage mat that gives the cat a mild electric shock. Elsewhere in this book you will find chapters on urination and defecation outside the litter tray, urine-spraying, and aggression to people and other animals.

SUMMARY

Keeping cats permanently inside requires more effort on your part. However the more you do for your cat—the more it will reward you. Many clients have told me that this program is lots of fun—and their cats like it too.

ADDITIONAL READING

Chamove, A.S.: "Environmental Enrichment: a review ," *Animal Technology*, 1989.

Holmes R.J.: "Environmental Enrichment for Confined Dogs & Cats.", *Animal Behaviour, Proceedings*, 1993. Post Graduate Committee in Veterinary Science, University of Sydney (NSW, Australia)

Landsberg G.M.: "Feline Scratching and Destruction and the Effects of Declawing." *Veterinary Clinics of North America: Small Animal Practice*, 1991.

Tudge C.: "A Wild Time at the Zoo," *New Scientist*, 1991.

Turner D.C., Bateson P., editors: *The Domestic Cat: The Biology of Its Behaviour* Cambridge University Press, Cambridge (UK), 1988.

Wemelsfelder F.: The Concept of Animal Boredom and its Relationship to Stereotyped Behaviour. Ch. 4 in Lawrence A.B., Rushen, J. (editors), *Stereotypic Animal Behaviour* CAB International, Wallingford, 1993.

INDEX

Page numbers in **boldface** refer to illustrations.

T.F.H. offers the most comprehensive selection of books dealing with cats. A selection of significant titles is presented below; they and many other works are available from your local pet shop.

TS-173, 304 pages
over 400 full-color photos

TS-127, 384 pages
over 350 full-color photos

TS-152,
448 pages
over 400 full-color photos

TW-103,
256 pages
over 200 full-color photos

TS-219,
192 pages
over 90 full-color photos